HEXAPOD STORIES

昆虫的故事

[美] 伊迪斯 · M. 派奇 著　亢海宏　洪梅 译

天津出版传媒集团
百花文艺出版社

图书在版编目（CIP）数据

昆虫的故事 /（美）伊迪斯·M.派奇著；亢海宏，洪梅译. -- 天津：百花文艺出版社，2016.7(2022.1 重印)
ISBN 978-7-5306-6913-6

Ⅰ.①昆… Ⅱ.①伊… ②亢… ③洪… Ⅲ.①昆虫-儿童读物 Ⅳ.①Q96-49

中国版本图书馆 CIP 数据核字(2016)第 173223 号

昆虫的故事
KUNCHONG DE GUSHI
(美) 伊迪斯·M.派奇 著

出 版 人：薛印胜
选题策划：杨进刚　　美术编辑：郭亚红
责任编辑：李 跃　　封面设计：明轩文化 TEL:23674746·王 烨
出版发行：百花文艺出版社(天津)有限公司
地址：天津市和平区西康路 35 号　邮编：300051
电话传真：+86-22-23332651（发行部）
+86-22-23332656（总编室）
+86-22-23332478（邮购部）
网址：http://www.baihuawenyi.com
印刷：三河市同力彩印有限公司
开本：880×1230 毫米　1/32
字数：128 千字
印张：9.5
版次：2016 年7 月第1 版
印次：2022 年1 月第3 次印刷
定价：39.80元

如有印装质量问题，请与三河市同力彩印有限公司联系调换
地址：河北省三河市黄庄镇赵家庄同力彩印
电话：17710717619

To Introduce Twelve Little Hexapods
让我们来认识十二只六脚节足小昆虫吧

The Hexapods are funny folk who have six feet. That is they have six when they are grown up, though some of the children have none at all, and some have as many as twenty-two. You can tell from this that they are strange people, and you may call them fairies if you like!

They have wings,—the grown-up ones do,—wonderful wings of many shapes and colors. Luna's wings are green,—pale, pale green,—and very lovely, with a purple border on them. Perhaps there is nothing more beautiful in the world than Luna's wings. When Van flies, you can see the yellow edge of her brown

Van was ready to enjoy a New Year's dinner.
蝴蝶帆儿准备享用一顿新年大餐。

wings; and when she alights—hesto! presto! you can see nothing at all; for she disappears from sight even though she is near enough to touch. Carol wears her wings neatly folded like a fan, except when she is using them. And Gryl, the little black minstrel—oh, Gryl fiddles with his wings.

They do queer things that we could not do if we tried. Old Bumble sleeps for more than half a year, and then wakes, thinking nothing of it at all, as if that were the most natural way to take a nap. Keti starts to build himself a log cabin before he is a day old; and finishes it, too, in time, with no one to show him how. And Cecid bewitches the willow with a magic no one else can learn.

Yes, you may call the Hexapods fairies, if you like; but you must never, never forget that they are every bit as real and true as you are, even if they are so very different.

They are not far away, not farther than the flowers or the trees or the nearest brook. And there are so many millions of them that every child in the world might have some for pets and they would never be missed.

And let me tell you this, for this is very important: although Hexapods are common and easy to find, there is not one among them all that does not have a story

about his life so strange and interesting that he is worth watching just to find out what his story is.

Are you pleased to know that, whether you are in the country or in the city, and whether it is summer or winter, you are living right in the midst of Hexapod Land, where you have these most wonderful fairies for next-door neighbors?

EDITH M. PATCH

六脚节足小昆虫一家可有意思了。他们有六只脚，确切地说，他们长大后有六只脚。可奇怪的是他们家有些孩子根本就没有脚，而有些孩子竟然能长二十二只脚！所以，如果你乐意，就叫他们“小精灵”好了。

他们有翅膀，是长大后才有的。这些翅膀奇形怪状、绚丽多彩，好看极了。月形天蚕蛾卢娜的翅膀是绿色的，淡淡的绿，镶着紫色的边，非常漂亮，说不定是世界上最漂亮的东西呢。蝴蝶帆儿在空中翩翩起舞，棕色翅膀的黄边清晰耀眼一览无余；不过，她落下来的时候，眨眼之间就不见了，即使在你的眼皮底下，她也会消失得无影无踪。蚱蜢欢欢不用翅膀的时候，总是把它们整整齐齐地叠起来，像把扇子似的背在身上。蟋蟀唧唧，噢！这个飞来飞去的小乐手！黑色的翅膀就是它的小提琴！

他们的本领稀奇古怪，让我们望尘莫及。黄蜂老奔奔一睡就睡大半年，醒来后还心安理得，觉得只不过是打了一个盹而已；袋袋蛾凯弟出生还不到一天就开始搭建自己的小木屋，不仅能如期完工，而且完全是无师自通；锯蜂塞西德·西德·多米会给柳树施魔法，这种魔法我们谁也学不会。

不错，如果你乐意，你可以把这些六脚小昆虫叫作“小精灵”。但是，千万千万要记住：尽管他们奇特无比，但他们一切的一切就和你一样是真实存在的。

他们近在咫尺，就像我们身边的鲜花、绿树和附近的那条小溪一样。他们的数量如此庞大，有数百万之众，即使世界上的每个孩子都捉一些回来当宠物，也不会觉得少了谁。

让我告诉你一件事情，因为这件事情很重要：尽管六脚节足小昆虫随处可见，可是，他们每只身上都有一个又神奇又有趣的故事，所以，哪怕是为了发现这些故事，他们也值得我们去观察。

无论在乡村还是在城市，也无论是盛夏还是寒冬，你周围到处是六脚节足小昆虫，这些奇妙的小精灵就住在你家隔壁。知道了这些，你是不是很开心呀？

伊迪斯·M.派奇

目录

1

Van, the Sleepy Butterfly Who Was Wakened by A January Thaw

冰雪消融的一月，贪睡的蝴蝶帆儿醒来了

Van was having a happy time making her New Year's calls.

She had crept out of bed about noon, for there was no cold wind blowing, and the sun had thrown its warm rays against the loose piece of bark under which she was sleeping. These warm rays had wakened her. That was a pleasant way of starting January, to have the sun knock at her door with its own sort of "Happy New Year! "

过年了，帆儿走亲访友，过得好开心呀！

她从床上爬起来的时候大约是中午时分。那时寒

风已停止了呼啸，太阳暖洋洋地照着她身上的树皮，正是这些温暖的阳光叫醒了她。一月份刚开始，就有太阳来敲她家的门，以特有的方式对她说“新年好！”，那可真是惬意呀！

Other pleasant things happened, too. To begin with, Van was thirsty; and there is much to enjoy in being hungry and thirsty if there is food and drink near by. She had had nothing to eat during November or December, as it had been too cold for two months to do anything but sleep; and she was ready now to enjoy a New Year's dinner.

还有别的高兴事呢。首先，帆儿很渴。要知道一个人又饿又渴的时候，刚好附近有吃有喝，那才叫享受呢。十一月和十二月的时候，她没东西可吃。那两个月太冷，除了睡觉，什么也不能干。现在好了，她已经做好准备，打算好好享用一顿新年大餐了。

So she must have been glad when she had some invitations to "come and have a feast." Of course these were not written in notes, and put into envelopes, and

stamped, and brought her by Uncle Sam's post-man. Nor were they brought her by messenger boys who said in words, "You are invited to dine at Mrs. Appleby's." Those would have been silly ways, for Van was a butterfly, and what did she know about written or spoken invitations or dinner-bells?

因此，收到那些让她“快来美餐一顿”的邀请，她一定高兴极了。当然，这些邀请函不是写在纸上、装在信封里、贴着邮票由萨姆叔叔的邮差送来的，也不是当差的小男孩跑过来对她说“苹果拜（applepby: applepie 的谐音，这里表示传信的小男孩口齿不清）夫人请你去她家吃饭”。这些方法都太傻了，因为帆儿是蝴蝶，她哪里懂什么书面邀请、口头邀请或者晚餐铃声呀？

But there was a way for all that to let her know the table was spread for her. And after all, what better way, when folk (either boys or butterflies) have been without food until they are hungry, than to be tempted by good smells?

但是大家自有办法让她知道餐桌已经为她摆好了。毕竟，人们（孩子也好蝴蝶也罢）好久不吃东西，已经饥肠辘辘的时候，有什么比香味更有诱惑力呢？

The first invitation that the air brought Van was from an old apple tree that lived near the edge of the woods. My, my! What a good-smelling one it was! The apples had frozen on the ground, and now they had thawed and were soft, and the juice was like cider. Oh, oh! What a feast for a thirsty butterfly!

空气给帆儿捎来的第一个请帖是树林边一棵老苹果树发出的。天哪,天哪!真香呀!冬天,这些苹果在地上冻硬了,这会儿解冻了,变软了,汁液就像苹果酒一样芳香浓郁。哦!哈哈!嗓子冒烟的蝴蝶这下可要喝个够了!

Other guests had been before her. Brother Rabbit had eaten there more than once. Mr. Fieldmouse had nibbled up half an apple and left a pile of crumbs in its place. And the birds? There was one that very minute pecking with its thick short bill at an apple that had caught in the branches. So bird and beast and butterfly were all made happy by the New Year's gift of the old apple tree.

在她之前别的客人就已经来过了。兔子兄弟已经不止一次在这里用餐;田鼠先生啃掉了半个苹果,在地上留下了一堆渣子;鸟儿呢?此时此刻,一只鸟儿伸着又粗又短的嘴巴正在那儿啄卡在树枝上的苹果呢。可见，老苹果树的新年礼物让鸟儿、兽儿和蝴蝶们都感到十分快乐。

The second invitation the air brought Van was from a maple tree that lived down the same lane not far away. This was such a good one that Van could hardly get there fast enough. She fluttered her wings in a hurry to get a quick start, and then sailed for a little way. Then she fluttered her wings again as if she wanted to get there before it was too late. And no wonder! For there was a broken branch on the maple and a little icicle where some sap had dripped out and had frozen, and now the icicle was melting. Maple syrup! Oh! oh! oh!

空气给帆儿捎来的第二个请帖是一棵枫树发出的，也在不远处的那条路边。这次的请帖太香了，她恨不得马上就能赶过去赴宴。她扇起翅膀匆匆忙忙就上路了，然后她只滑翔了一小段就又扇动翅膀直奔宴会而去，生怕去晚了。这也难怪！原来这棵枫树的一根树枝断了，滴下来的枫树汁冻成了冰柱，这会儿冰柱开始融化了。枫树汁！太香了！太香了！太香了！

The third invitation that the air brought to Van was—But we shall never know because, just as she was starting to her third feast, a boy and a girl came racing down the lane.

空气捎给帆儿的第三个请帖是……唉！我们永远也不得而知了，因为她正要出发去赴宴的时候，一个小男孩和一个小女孩从路那边跑了过来。

"Oh, see! " they called; "a butterfly! A big beauty! Just think! A butterfly on New Year's Day! Let's take it home to show! "

"呀，快看！"他们喊道，"蝴蝶！一只大蝴蝶！好漂亮啊！想想看！新年第一天我们有只蝴蝶该多好！我们把她逮回家让大家看看吧！"

So they ran after Van, who was spreading her brown wings with yellow borders for a slow sail.

于是，他们跟在帆儿后面追了起来，此刻，帆儿正张着她那镶着黄边的棕色翅膀慢速滑翔。

Oho! catch Van? Why they couldn't even see her! What had been a large butterfly, with wide showy wings, a minute before, now looked like a ragged bit of bark on a tree near by. Van had hidden. And she was almost near enough to touch, though perhaps a little too high. She had hidden right in plain sight. And all she had done was sit on the bark of the tree and fold her wings above her back and keep still. Catch Van? Why Van could fool a bird in a game of hide-and-seek !

噢喔！逮住帆儿？这会儿他们连看也看不到她了！一分钟之前那只翅膀艳丽的大蝴蝶，这会儿看上去跟旁边树上一块粗糙的树皮差不多。帆儿藏起来了。其实她就在他们触手可及的地方，不过也许稍微高了一点点；她就在他们的眼皮底下，只不过是把翅膀叠起来搭在后背上，一动也不动地坐在树皮上罢了。想逮住帆儿？可没那么容易！即便是和鸟儿捉迷藏的时候，她都能把人家捉弄得团团转呢。

By the time the children gave up the hunt, the sun was under a cloud and the wind felt chilly. So Van did not have any more feasting that day, or make any more New Year's calls. She was near her shelter of loose

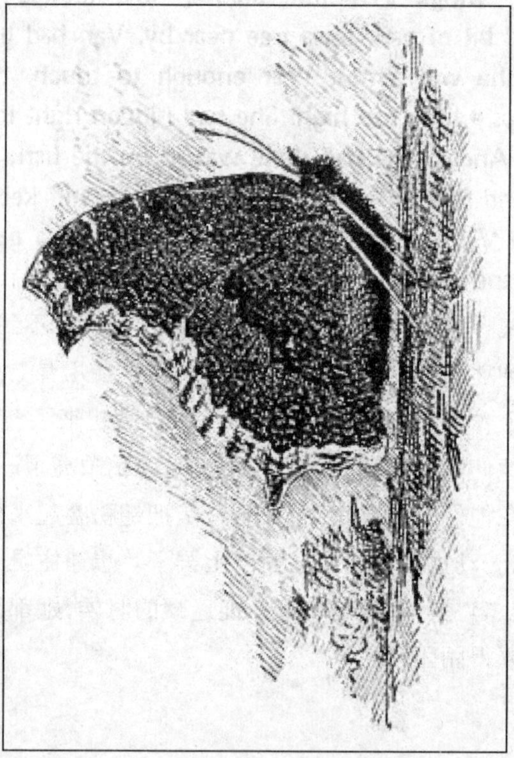

Van now looked like a ragged bit of bark.
帆儿现在看起来就像一块儿糙树皮。

bark and crept up under cover out of the wind. She had had a good time. She had had cider and maple-sap enough to last her until the next thaw spell —be it one week or two, or one month, or two, or even three. Time was all alike to her when the weather was cold. —Short or long, it was the same! She just slept! You may ask the wisest man you know why it did Van no harm when the weather was cold as zero. If he is very wise, indeed, he will tell you, "I do not know why."

孩子们停止追逐的时候，太阳已经躲到了云背后，风也感觉凉飕飕的。所以，那天帆儿没再去赴宴，也没再去走亲访友。为她遮风避雨的那块松松的树皮就在附近，于是，她爬上去钻到里面去避风了。这一天她已经过得很开心了。她享用了苹果酒和枫树汁，足以让她坚持到下一次冰雪融化的时候——不管是一个礼拜还是两个礼拜，也不管是一个月还是两个月，哪怕是三个月也没关系。天冷的时候，时间长短对她来说没什么区别，都一样。她只管睡觉！你可以去问问你认识的最聪明的人，为什么寒冷如冰的天气不会冻坏帆儿？要是他真的很聪明的话，他会告诉你：“我也不知道这是为什么。”

But our not knowing why made no difference to Van. She did live through the cold winter in the north. With no shelter but a loose bit of bark, Van now looked like a ragged bit of bark, just as her grandmother had lived one winter in a hollow tree, and her great-great-grandmother had lived under the roof of an old empty shed.

不过,对帆儿来说,这又有何妨呢?因为,她的的确确度过了北方的天寒地冻。现在的帆儿,除了一块松落的树皮之外,一无所有,看上去她自已也像一块粗糙的树皮。她的外婆当年冬天住在一棵空心树里的时候,不也是这副模样吗?她的外婆的外婆的外婆当初住在一个破空棚的棚檐下的时候又何尝不是这副模样呢?

As spring came on, Van had invitations from the pussy willow, where she met some pretty flies with stripes on their bodies. She called, too, on the trailing arbutus, where she met Old Bumble more than once; for they both drank from those sweet pink cups and carried pollen from flower to flower.

春天来了,帆儿收到了银柳的邀请,在那里她见到一些苍蝇,他们身上长着条纹,很漂亮。她还去了野草莓那里,在那里,她不止一次见到了老奔奔,因为他们俩都去那些香甜的粉色花蕊里酣饮，也都在花丛里搬运花粉。

Some time in May Van found that she could not spend all her time in calling on flowers, however much she liked their nectar, and however much they needed to have their pollen carried for them. So she hunted for a willow tree, and made a little ring around one of its twigs. This ring was set with fifty jewels, the very best she had to offer to the world. Of course, these jewels were her eggs.

五月里有一段时间，帆儿发现，无论她有多么喜欢那些琼浆玉液，也无论那些花儿多么需要有人帮忙运花粉，她都无法整天待在花丛里了。于是，她找了一棵柳树，在一条柳枝上做了个小戒指，还在戒指上镶了五十颗宝石，这是她倾尽所有献给这个世界的最好礼物。没错，这些宝石就是她产下的卵。

In about two weeks they began to hatch, and Sister Essa was the very first of them all to bite around the edge of her egg-shell, until the top lifted like a little lid and out she came, looking much too long to be curled up in the shell she

crept out of.

大约两周后,小宝宝就开始咬破卵壳出来了。第一个出来的是艾莎姐姐。她沿着卵壳咬呀咬,直到头顶的卵壳像小盖子一样打开,她就出来了。她的身子看上去很长,怎么也不像是刚刚从那个卵壳里爬出来的。

Essa did not go away from her brother and sister caterpillars. When they were all hatched, they crept off together and lay in a row side by side, with their heads at the edge of the leaf. There they had their breakfast, which it took them several days to eat.

艾莎并没有离开她的毛毛虫弟弟妹妹们。等大家都从卵壳里出来之后，他们才一起爬到柳叶边，头朝外,躺成一排。然后,他们开始吃早餐,这顿早餐他们要吃好几天。

And what do you suppose it was? No, their mother did not bring them cider from old soft apples, or syrup from broken maple twigs, or nectar from flowers. She was not like Old Bumble, who fed her babies every day. Van did not bring her fifty children

one single thing to eat. Sister Essa was hungry, too; and so were the rest of the family. And here they were left on a willow branch, where there was nothing at all but leaves — food their mother never could have eaten with her long tongue, if she had uncoiled it and tried.

你猜他们吃的是什么？不对，他们的妈妈没有从烂掉的苹果那里给他们带来苹果酒，也没有从折断的枫树那里给他们带来枫树汁，更没有从花儿那里给他们带来花蜜。她可不会像黄蜂老奔奔那样每天都喂自己的小宝宝。帆儿没有给她的五十个孩子带回一丁点儿吃的。艾莎很饿，弟弟妹妹们也都很饿。就这样，他们被丢在柳树枝上，除了吃柳叶以外，别无选择——要是他们的妈妈曾经展开这些柳叶尝上一尝的话，那她今后绝对不会再用她的长舌头去碰这种食物的。

But we needn't worry about those babies. In less than the flick of a minute Sister Essa had nibbled a tiny green bite out of the top of the willow leaf, nodding her little head over it as her jaws opened and shut as if there were nothing better to be had for a breakfast. All her brothers and sisters were nodding their heads just the same way, and while they were about it, they nibbled off the whole top of the leaf just as if it were a

green layer cake and they wanted only the shiny frosting. Before breakfast was over, Sister Essa led them to another leaf, where they lay side by side in a row as before, and ate until their skins were too tight to hold another bit of shiny green frosting.

不过，我们也不用为这些小宝宝担心。因为，眨眼之间，艾莎姐姐就把绿油油的柳叶尖啃掉了一小块，只见她冲着柳叶低着小脑袋，嘴巴一张一合吃得那么津津有味，似乎再也没有比这个更好吃的早餐了。她的弟弟妹妹们也都像她那样低着小脑袋吃得全神贯注。他们把柳叶外面那层整个儿地都啃掉，就好像是在吃一个夹心蛋糕，上面那亮晶晶的糖霜才是他们的最爱。这边还没吃完，艾莎姐姐就把大家带到另外一片柳叶上，他们还像刚才那样躺成一排，专拣那绿油油的糖霜吃。他们吃呀、吃呀，一直吃到肚皮鼓鼓的，再也吃不下为止。

That was a sign that their breakfast was over; so they spun a thin mat with silk which they spilled out of a little tube near their lower lips, and took a nap on the mat.

这时早餐该结束了。他们的下嘴唇附近有一个小管子，里面可以吐出丝线来。于是，他们就用这些丝线织成小毯子，然后，就躺在小毯子上打起盹儿来。

The first day they rested quietly, but the second day they acted as if they were having bad dreams and tossed their heads a great deal. In fact Sister Essa jerked so hard at last that her little skull came off like an empty shell. By that time she was wide-awake, and crept out of her tight skin through the collar-hole the skull left when it tumbled off. Before she had time to turn around, all her brothers and sisters were jerking their skulls off, too, and creeping out of their skins through the collar-hole.

第一天他们睡得很安静，可是第二天他们不停地摇头晃脑，好像是在做噩梦。结果，艾莎姐姐因为用力过猛，脑壳竟然像空壳一样掉了下来，留下一个像领口那样的洞。这个时候她彻底醒了，于是，她从那个洞口里爬了出来，蜕掉了身上那紧绷绷的皮。她还没来得及转身，弟弟妹妹们也都跟她一样，甩掉脑壳，从领口一样的洞里爬了出来，蜕掉了身上的皮。

Something funny had happened to them and they never looked the same again. They now had new

heads, with bigger jaws and fine new stretchy skins.

他们已经发生了有趣的变化，再也不是以前那副模样了。现在，他们长出了新脑袋，嘴巴也变大了，而且连皮肤都换成了新的，很有弹性。

After that nap that had had such a queer end, they were hungry; so they went off to some new leaves (this time one was not big enough to hold them all) and lay in rows and ate their luncheon. It was so good they did not stop for nearly a week. When their luncheon was over at last, they spun another thin silk mat and had another nap. They woke in about two days, jerking their skulls off again, and crept through the collar hole in their skins just as they had before. They now had still bigger heads, and skins that were stretchier than ever.

从这个结局古怪的睡梦中醒来之后，他们饿了。于是他们离开以前的叶子爬到了新叶子上躺成一排开始吃午餐，不过，这次一片叶子已经不够他们躺了。这顿午餐太好吃了，他们不停地吃，一连吃了快一个礼拜。终于吃完午餐之后，他们又织了一块丝毯，躺在上面又睡了一觉。大约过了两天，他们醒了，跟上次一样，又晃掉了脑壳，从外皮上像领口一样的洞里爬出来。现在他们的头更大了，皮肤也更有弹性了。

Well, that was the way Sister Essa went on doing, until she had had her dinner and another luncheon and her supper. Each meal lasted several days, with a day or two for a nap in between. Every time she wakened, she pushed and jerked inside her old skin, until her skull fell off like an empty nut-shell; and when she crept out of the collar-hole she looked different from the way she looked when she went to sleep.

艾沙姐姐就这样吃了睡，睡了变，变了吃。每顿饭都要吃上好几天，吃完之后总要睡上一天或者两天，每次睡醒之后，都在原来的老皮肤套里摇来晃去，直到脑壳像个空坚果壳一样掉下来。每当她从领口一样的洞里爬出来的时候，模样就已经跟睡觉前不一样了。

By the time she was eating her supper, her skin was a soft black color, with little white specks like a "pepper-and-salt " suit. Down the middle of her back was a row of pretty red spots, and growing all over her sides were black spines with pointed branches. She was now two inches long and a fine-looking caterpillar—after one got used to seeing her.

到吃晚餐的时候，她的皮肤已经变成了炭黑色，上面布满了小白点，好像是芝麻呢套装。沿着她的脊背中线有一排漂亮的红点，身体两侧则长满了尖尖的黑色绒刺。她现在已经两英寸长了，看惯了之后，还是一条漂亮的毛毛虫呢。

As her brothers and sisters had all had the same sort of time growing that she had had, they were fine-looking, too, and so big that the fifty of them together made the tip of the willow branch hang down. They ate more for this meal than for any other, and they did not nibble just the shiny frosting as they did at breakfast when they were tiny—They gobbled up all the flat green cakes on the branch. This would have been a bad thing for a little tree which needed all its leaves to grow with. Of course, if they had been on a small tree, it would have been better to take them off. But this was a big, big one, forty years old, and it was growing wild near a brook, with no gardener to trim off some of its branches; and what leaves Essa and the others ate could be spared as well as not. Of course, their supper must be a hearty one, for it had to last them until they were

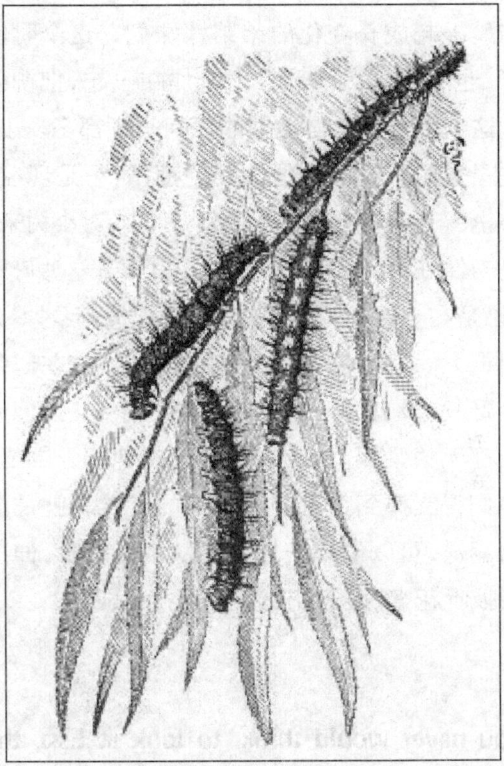

Each meal lasted several days.
每吃一顿能管好几天。

butterflies, like Van, with a long tongue to uncoil when calling on the flowers to sip nectar and to carry pollen.

和她一起成长变化的弟弟妹妹们，这个时候也变漂亮了，也长大了。他们五十个加在一起，把柳条都压弯了。这顿饭他们吃得比别的任何一顿都多。小时候，他们吃早餐时只吃那些绿油油的糖霜，可是现在，他们把柳条上所有的柳叶蛋糕都吞了下去。对小树来说，这可不是一件好事，因为小树的生长离不开这些叶子。当然，如果是在一棵小树上，完全可以把他们拿掉。可是，这棵树很大、很大，都有四十年树龄了。它在一条小溪旁野生野长，没有园丁来帮它修剪枝条，艾莎他们吃多少树叶都无关紧要。当然，这顿晚餐他们一定要多吃，因为他们得支撑到自己变成蝴蝶的时候，就是像帆儿那样的蝴蝶，可以像她一样去花丛中做客时，伸出长长的舌头吸吮花蜜、搬运花粉。

You never would think, to look at Essa, that she would ever fly; for there she sat clinging to the branch with ten of her feet and drawing the edge of the leaf down to her mouth with her other six feet, and she didn't have a sign of wings anywhere on her back.

看艾莎的样子，你绝不会想到她竟然会飞。她坐在那里，十只脚勾着树枝，另外的六只脚拽着柳叶的边往嘴里送，脊背上连一点儿翅膀的影子都看不到。

Ah, but Essa could do several things you would never think she could! She had never done them before—Why should she now? You might not think she could creep head first down the trunk of the tree, and take a walk, as fast as she could hurry, along the ground, until she came to an old fence; and climb the fence, and spin a silk peg on the lower edge of a board, and fasten her hind-legs to the silk peg, and let go with all her other legs, and hang there head down until her skull split and her skin ripped down the back seam!

可是，啊！艾沙有好几个你绝对想不到的本事！以前她从来没露过这些本事，为什么现在却要显露出来呢？想不到吧？她会头冲下，顺着树干爬到地上，然后，拼尽全力急急忙忙走到一根旧栅栏跟前，爬上去，在栅栏板的下边吐出丝织成一个丝栓子，然后，后腿粘住这个丝栓子，松开其他腿，整个儿倒挂在那里，直到脑壳开裂，外套沿着背缝裂开脱落下来！

You wouldn't know how Essa could do that, would you? And if you ask the wisest man you see how a caterpillar can do wonderful things like that just once in her life, without learning or without any one to show her about making a silk peg, maybe he will tell you he doesn't know, either.

你不知道艾莎这个本事是怎么来的吧？为什么毛毛虫一辈子只显示一次这样了不起的本事呢？她可从来也没学过怎么做丝栓子，也根本没人教过她怎么做呀。要是你去问你见过的最有智慧的人，大概他也会说不知道的。

But our not knowing how she can do it made no difference to Essa. That is just what she did when she had finished her supper. and while she was about it, she changed into a chrysalis, which looked no more like a butterfly than it did like a caterpillar.

不过，对艾沙来说，我们知不知道她怎么会有这样的本事又有什么关系呢？吃完晚餐后她就这样了。一边吃，就一边变成了一个蛹，看上去不像一只蝴蝶，倒更像一只毛毛虫。

Well, there they hung, Essa and her forty-nine funny brother and sister chrysalids, for about ten days; and none of them knew anything about Lampy's fireworks on the Fourth of July. Soon after that Essa broke her chrysalis case, and tumbled out head first. She didn't tumble far, for her feet caught hold of the empty case, and she hung there with her soft little wings down, until they grew big and stopped throbbing. She clung with four feet; but you must not think she had no more feet than a cat or a dog, for her first pair were folded against her breast and covered up by her pretty brown fur.

就这样，艾莎和她的四十九个蝶蛹弟妹们滑稽地倒挂了十天左右，像一串小灯笼，其实，他们谁也不知道七月四日挂灯笼、放烟火这回事。（译者注：七月四日是美国独立日，人们悬挂起灯笼、燃放起烟花以示庆贺。Lampy就是lamp。）没过多久，艾莎就弄破了蛹壳，先把头伸了出来，不过，没伸出多远，因为她的脚还抓着空空的蛹壳。她倒挂在那里，耷拉着软软的小翅膀。渐渐地，翅膀变大了，停止了扑打。她用四只脚挂住身子，可是，你千万不要以为她像小猫或小狗一样只有四只脚。她最前面的那双脚叠在胸前，被她那漂亮的褐色绒毛遮住了。

Some people said Essa's wings were purple, and

There they hung on their silk pegs.
他们吊在自己的丝钩上。

some said they were brown. I don't know what you would think about their color. But all agreed that the border was pale yellow, and that next the yellow on the upper side was a dark band with lovely pale blue spots on it. And everybody who saw her said that she was a very beautiful butterfly when she spread her wings open, and that, when she folded them shut above her back, she looked like a piece of bark.

有人说艾莎的翅膀是紫色的，有人说是褐色的，我不知道你怎么看。不过，大家都认为她的翅边是浅黄色的；紧挨着黄色的边，上半面是一条深色的带子，上面撒满了有趣的浅蓝色斑点。所有见过她的人都说，展开翅膀时她是一只非常美丽的蝴蝶，可是当她把翅膀收起来拢在背上的时候，她看起来就像一块树皮。

Essa flew about and made calls on the flowers, for there was flower nectar instead of frozen apple-cider in July. After a time she made a ring about the twig of a tree all set with her egg-jewels. I have forgotten whether she chose a willow or an elm; but it made no difference which, for her daughter Opie, when she hatched, could eat either one. If it had been an oak, Opie would have starved to death; but of course Essa

would not have left her eggs on an oak or a pine tree.

艾莎飞来飞去，到花儿那里去做客，因为，七月份没有冰冻的苹果酒了，只有花蜜。过了一段时间，她在一棵树的树枝上了做了一个戒指，上面镶着她那像珍珠一样的卵。我忘了她选的是柳树还是榆树了，不过，这没关系，因为无论哪个，她的女儿奥培孵化出来后都能吃。假如她选的是橡树，奥培早就饿死了。当然，艾莎是不会把卵产在橡树或松树上的。

Well, Opie and her brother and sister caterpillars ate and napped and grew, and changed into chrysalids and then into butterflies, just as their mother and their uncles and aunts had done.

好了，奥培和她的毛毛虫弟妹们就像他们的妈妈、舅舅们和姨妈们那样，吃了睡，睡了吃，吃了长，然后变成蝶蛹，又变成蝴蝶。

But by this time it was the fall of the year, and Opie found her life much more like her grandmother's than it was like her mother's; for Essa had been a summer butterfly and Opie, like Van, was

a winter one. So she flew about and called on the fall flowers, and when the days grew cold, she found a shelter as her grandmother Van had done, and went to sleep, clinging with four feet to the roof of her winter bedroom and folding the other two close to her breast.

可是，这个时候已经入秋了，奥培发现自己的生活更像她外婆，而不像她妈妈，因为她的妈妈艾莎是夏天的蝴蝶，而奥培和她的外婆帆儿一样是冬天的蝴蝶。所以，她飞来飞去，到秋天的花儿那里去做客，天气变冷的时候，她像外婆当年那样，找了一个遮风避雨的地方当作过冬的小屋，钻进去，四只脚紧紧勾住屋顶，另外两只脚叠放在胸前，睡起了大觉。

There she rested all the winter long, except when the days were warm enough to thaw her out. For be it one month or three or even four, time was all alike to her when the weather was cold—Short or long, it was all the same. Opie just slept!

整个冬天她就待在这里足不出户，除非天气转暖，冰雪消融。天冷的时候，时间对她来说都一样——不管是一个月还是三个月，哪怕是四个月。长也好，短也好，都一样。奥培只管睡觉！

2

Old Bumble
黄蜂老奔奔

Candlemas Day was bright and fair.

Perhaps the ground-hog came out of his hole and was scared by his shadow and went back to bed again. I do not know.

But I know Old Bumble didn't come out of her hole that day. Her legs were all cramped up with the cold, and even her pretty black and yellow fur couldn't keep her warm.

圣烛节那天风和日丽。(Christmas 是圣诞节,这里的 Candlemas 由 candle 与 mas 和二为一，表示蜜蜂世界里的节日,也是指 12 月 25 日。)

也许,土拨鼠从洞里出来过,可是又被自己的影子吓到了,赶紧回去睡觉了。这个我不知道。

可是,我知道那天老奔奔没有出洞。她的腿冻僵了,连她那漂亮的黑黄相间的绒毛外套都不管用了。

You see, she had chosen to make her winter bedroom in a little cave on the north side of a dry bank, and the sunshine did not touch it. This did very well last August when she went to bed. If she had been in a sunny place then, it might have been too hot for her. It didn't take much to disturb her when she first went to sleep. If anything had got into her bedroom then and touched her, she would have shaken herself and gone away and found or made a new place. In August she was what is called "a very light sleeper" and couldn't bear to be touched.

你瞧，她选的过冬小屋在一个干燥的河堤北面,很难见到阳光。去年八月她进去睡觉的时候，还很不错。那个时候，如果她选的是阳光充足的地方，那就太热了。她刚入睡的时候,很容易惊醒。如果那个时候有什么东西进了她的卧室碰了她,她就会抖抖身子,离开那个地方,再找一处新居,或者再造一个新居。处在八月份的老奔奔,就是我们所说的那种“睡觉很轻的人”,受不了任何打扰。

In winter she was different. If the ground-hog had pawed her cave open and rolled her out of bed on Candlemas Day, she would not have known anything about it!

冬天的时候,她就不这样了。圣烛节那天,即使土拨鼠用爪子刨开她的洞,把她从床上拨拉下来,她也会毫无知觉。

He didn't, so there she slept, though she had been napping for six months already and would very likely keep at it for two or three more. Isn't that pretty lazy for a bee?

土拨鼠并没有这么做,所以,老奔奔还在那里睡着。她已经睡了六个月了,而且很有可能接着再睡两三个月。对蜂儿来说,这是不是懒了点?

She had not done one bit of work before she went to bed either—that is, nothing except to straighten out

Old Bumble's long nap.
老奔奔睡了个长长的午觉。

her bedroom a little; and as that was hardly more than a hole in the ground, the process did not take her long. She had not even hunted for her own dinner, which was to last her all winter. She helped herself to some fresh honey that her older sisters had put into a honey-mug, and drank enough to fill her honey-sack, and then went off and crept into bed. Pretty lazy for a bee, wasn't she?

另外，睡觉前，她一丁点儿活儿都没干，只不过稍稍拽了一下她的卧室而已。再说，她的卧室也就地上的一个小洞那么大点儿，所以，整个过程没花她多长时间。她甚至连过冬的晚餐都没有去找，而是毫不客气地喝了一些姐姐们放在蜜罐里的鲜蜜，喝得饱饱的，把自己的蜜袋都装满了，然后就爬上床去睡觉了。对蜂儿来说，这是不是懒了点？

She not only slept through Candlemas Day, but St. Valentine's Day came, with its pretty shower of cards and letters, and she didn't wake up then. George Washington's birthday found her sleeping still; and she didn't even dream while people were putting green ribbons in their buttonholes on St. Patrick's Day.

她不仅睡过了圣烛节，而且情人节那天漂亮的贺卡和信件纷至沓来，可是，她仍长睡不醒。乔治·华盛顿的生日那天（2 月 22 日）她还在酣睡。圣帕特里克节那天（3 月 17 日），人们在纽扣眼儿里别上绿色的丝带的时候，她睡得甚至连梦都没做一个。

It was not, I think, until April that she first roused herself and poked her sleepy head out of doors. Perhaps she was a little April Fool, for there was not much that she could do so early in the spring.

一直到四月份她才起床，睡眼惺忪地从门口探出头来。也许，她就是愚人节的那个“小愚人”，因为春天这个时候太早，她没有什么事情可做。

Maybe, though, she could find something to drink, to make her feel better after her nap of many months. Yes, the cool wind brought her a sweet smell from the trailing arbutus, the loveliest blossom that grows in the spring. For Old Bumble lived in a land where people had not yet robbed the woods of this dear flower, which used to be very common in the days of your grandfathers, and is getting to be so rare that, unless you and the rest of us are careful, there will

not be any left at all for your grandchildren to see.

不过,她也许可以找点喝的。睡了这么长时间,喝点东西会让她觉得舒服点。是的,凉风习习,给她带来了玻璃海棠香甜的味道,那是春天里盛开的最美丽的花儿。老奔奔住的那个地方,树林里还能见到这种珍贵的花儿,在你爷爷那个时代,这种花到处都是。可是,现在却越来越少见,如果我们大家不小心保护,它们就会完全消失,你的孙辈们也就看不到这些花了。

Well, Old Bumble found a few of the very earliest pink sprays in the sunny places, and she sang a happy humming song as she sipped at the tiny sweet cups.

好了,在阳光灿烂的地方,老奔奔找到了几枝最早开花的粉红色玻璃海棠,她一边唱着快乐的奔奔歌,一边吸吮着小花杯里香甜的汁液。

She did not stay up very long that day, for the sun soon went under a cloud and she felt like going to bed again. Then for a week or so it rained, so she took a nap until the weather was fine. After that, she got up and sipped from the different spring flowers as they

blossomed, but still slept through the colder days.

那天她没有熬很久，因为太阳很快就躲进了云层，她又想上床睡觉了。接下来大约一周时间都在下雨,所以,她一直睡到天晴才起来。这时,其他春天的花儿也陆续开放了,于是,她起身下床去缤纷的花丛里吸吮花蜜,不过天一变冷,她就还会整天大睡。

Did anyone ever tell you that a bee is a busy little thing?

Well, you see how Old Bumble spent the greater part of the year just dozing the time away. But wait and see what she did the rest of her days!

有人告诉过你蜂儿是勤劳的小东西吗？

唉,你都看到了,老奔奔一年里的大部分时间都是在睡梦中消磨掉的。不过,别着急,让我们看看接下来的日子里她都干了些什么吧!

As soon as the weather was warm enough she started out on a hunting trip. She buzzed slowly along near the ground, but this time it wasn't flowers she was after. She was house-hunting. Just bedrooms no longer

suited her. She was done with sleeping day after day. What she wanted now was a nursery. She must find a place where she could bring up a family of children.

天气刚一转暖,她就踏上了寻寻觅觅的征程。她紧贴地面慢慢地嗡嗡着向前飞去。这次花儿不是她的目标。她是在找房子。那几间卧室已经不够她用了。她已经不需要天天睡觉了。现在她需要的是一个育儿室。她必须找个地方抚养一大堆孩子。

So here and there, and there and here, she flew, singing her slow hunting song as she went. Now and then she stopped and peeped into a hole, to see if she liked it; and if it was not good enough for her home, she came out and hunted still farther.

所以,从东到西,从西到东,她飞呀飞,一边飞一边唱着她那舒缓的寻觅之歌。她时不时地停下来,飞到一个洞里仔细察看,看自己是不是喜欢;要是觉得不够好,她就出来,继续往前去找。

At last she found a place that would do nicely for her nursery. It was a home a field-mouse had lived in

once upon a time, and a field-mouse has very good taste about underground houses. Any way, Old Bumble liked the same kind.

终于,她找到了一个地方,很适合作她的育儿室。那里曾经住过一只田鼠,田鼠的地下房屋鉴赏力不错。不管怎么说,老奔奔也喜欢这样的房子。

Of course it needed tidying up a bit after being empty so long. So she went right to housecleaning as if she knew all about it. She had never done such a thing in her life before, but she was not so stupid that she had to be shown how to do everything. She had a way of getting things right the first time she tried. She had saved her strength for many months, and now she was going to use it.

当然,房子空了这么久,得好好收拾一下。于是她立刻开始动手打扫屋子,似乎很在行的样子。以前,她从来没有收拾过屋子,可是,她还不至于笨到事事都需要别人教的程度。就算是从来没做过的事情,她也有本事做得井井有条。她已经攒了好几个月的力气,现在,就要派上用场了。

How Bumble's nest looked inside?
老奔奔的巢里面看起来怎样?

Her nursery must be just exactly right! For one thing, it should be dry; and this empty mouse-hole had grown damp. So she worked about in the part she was going to use, and dried it out with the warmth of her body. She found the softest bits the mother mouse had left there, and shook them up with her jaws and piled them in a heap. Right in the very middle of this she hollowed out a little room, which covered her up, top and all, except a hole at one side which she kept for a doorway into her snug little nursery.

她的育儿室必须符合要求,一点儿都马虎不得!首先,应该干燥。这个空置已久的田鼠洞已经发潮了。所以,她从要用的那块地方着手,先用自己的体温把那里暖干。然后,她把田鼠妈妈留下的最软的绒毛捡起来,用嘴巴把它们摇松弄匀堆成一堆,在这堆绒毛的最中间挖出一个小屋,这小屋可以把她从上到下遮了个严严实实,只有一边留了一个口,可以通到她这间舒适温暖的小育儿室。

Then off she flew to the flowers for yellow pollen, which she gathered and packed into her pollen baskets on her hind legs. She brought back her load and put a lump of this yellow stuff, made moist and sweet with

honey, right on the floor of her nursery. Next she brushed some wax off her body and made a little nest of it big enough for a few eggs.

接下来,她飞出去采黄色的花粉。她的后腿上有装花粉的篮子，她把采集到的花粉塞进去，给它们掺上蜜,做成湿润香甜的一块黄疙瘩。她把这个黄疙瘩背回家,直接放在育儿室的地上,然后,从身上刷下来一些蜡,用这个黄疙瘩做成一个够放几只卵的小蜂巢。

Before night came on, she brushed off some more wax, and this she made into a honey-mug just inside her doorway, and into this she put what honey she had had time to gather.

趁着夜幕还未降临,她又刷下来一些蜡,比上次的多,在门后做成一个蜜杯,把以前一有空就去采的蜜都统统装了进去。

But she must not fly too late, for her eggs must not get cold. If they did, it would take them longer to hatch, and she needed her daughters to help her as soon as could be. She sat on the little wax nest to keep

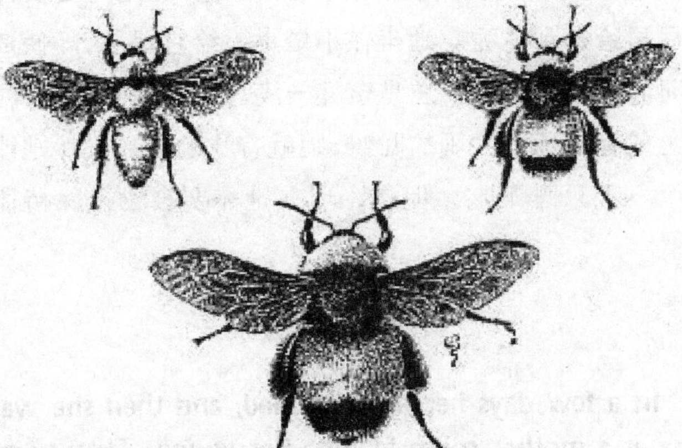

Old Bumble and a son and daughter bee.
老奔奔和他一儿一女的蜂宝宝。

it warm, and left it only long enough to fill her honey-mug, so that she could eat from it in the night or on stormy days. For, though she could sleep all winter without eating, she needed food now to give her strength.

但是她一定不能太晚飞出去，因为她的卵绝不能变凉。如果变凉，就要花更长的时间孵化，而她还要女儿们尽早给她帮忙呢。她坐在小蜡巢上给它保温，即使离开时间也不会很长，总是蜜杯一满马上就回来，这样，晚上或者刮风下大雨的时候，她就有吃的了。虽然她能整个冬天只睡不吃，但是，现在她需要食物来保持体力。

In a few days her eggs hatched, and then she was busy as a mother robin feeding her young. They were white little babies without hair or legs, and you never could guess to look at one that it would some day be a black-and-yellow furry bee. You could not take a peep at them, though, as you can at little birds, because their nest had a tight wax cover, and the nest was in the nursery, and the nursery was in a hole, and the hole was in the ground.

几天之后，她的卵就孵化了，她就像知更鸟妈妈一样，整天忙着喂孩子。她的小宝宝白白的，没有头发也没有腿。看其中随便哪一个的样子，你也绝对想不到有一天，它会变成一只蜜蜂，身体黑黄相间，毛茸茸的。你能偷看小鸟宝宝，可是，却偷看不到蜜蜂宝宝，因为它们的小巢上面盖着一层严严实实的蜡，小巢外面是育儿室，育儿室外面又是一个洞，这个洞在地底下。

Now, how could Old Bumble feed her little ones if she kept them shut up tight in bag of wax? Well, for part of their food they ate up the pollen paste she kept bringing and sticking close against their nest.

可是，如果老奔奔总是把宝宝们严严实实地关在一个蜡口袋里，那她怎么给它们喂食呢？是这样的，老奔奔不停地把花粉糊带回来，抹在宝宝们的小屋上让它们吃。

But that was n't all they had to eat. Their mother mixed some pollen with honey until it was thin, and then bit a hole in the top of the nest and dropped it from her mouth. Every time she did this, she mended with wax the hole she made, for her babies had to be

kept shut up tight.

但是，宝宝们不光吃这个。它们的妈妈在花粉里掺上蜜，调稀，然后，在它们的小屋顶咬一个洞，把调制的蜜粉糊从嘴里吐进小屋里。每次喂完宝宝之后，她都会用蜡把自己咬的洞补好，把宝宝们严严实实地关起来。

You can see that, with getting pollen and nectar from the flowers, and making pollen-paste and honey for the children, and brushing wax from her body to use in keeping the nest and honey-mug mended, and storing honey in the honey-mug for nights and stormy days, the mother bee was not so lazy as she seemed at first. We might call her a very busy Old Bumble indeed. A happy one, too, humming her cheerful song as she flew about on her out-door errands.

你看，她要去花儿那里采集花粉和花蜜，要为孩子们调制花粉和蜜糊，要从身上刷蜡修补小屋和蜜杯，还要在蜜杯里储蜜，以备夜晚和风雨天之需，所以，这个蜜蜂妈妈并不像当初看上去那么懒。实际上，她称得上是一只繁忙的老奔奔，同时，也是一只快乐的老奔奔，因为她在外四处奔波时，总是一边飞，一边嗡嗡嗡地哼着欢快的小曲。

In eleven or twelve days the little ones stopped eating and each one spun a thin cocoon about herself, for she must still be covered up. This would seem to give their mother a chance to rest. But no! It only meant new duties. She must clean away the wax and pollen from the cocoons, and make new wax nests and lay more eggs. Besides, she must keep the cocoons warm. So she spread her furry body out as big as she could make it stretch, like a mother biddy covering her little ones under her feathers.

大约过了十一二天，小宝宝们不再吃东西了，都在自己身上织了一层薄薄的茧，因为它们还得把自己遮起来。这样，它们的妈妈好像可以趁机休息一下了。可是不能！这下她又有了新的责任。她必须把沾在茧上的蜡和花弄掉，做一些新蜡巢，产更多的卵。另外，因为要给这些茧保暖，她还要尽可能地张开毛茸茸的身体，像鸡妈妈那样把自己的小宝宝们揽在身下。

After about two weeks, or maybe it was not quite so long, the young bees began to bite holes in the caps of their cocoons, which might make you think of chickens pipping their shells. They had had enough of being shut up. They were now coming out into the

world. This time Old Bumble did not try to wax them in. She helped them out. She might well be glad to see her daughters, even though they were queer and feeble little things. Their fur was as wet as a chicken's down when it first comes out of the shell. Their legs were so weak they could hardly toddle over to the honey-mug for their breakfast. After they had eaten, they crept back and cuddled down under their mother, until their fur was dry and fluffy and they felt strong.

过了大约两个礼拜，也许没那么久，小蜜蜂们开始在它们的茧帽上咬洞，这也许会让你想起小鸡破壳而出的情景。它们已经被关够了，现在，它们要破茧而出来见识大千世界了。这次，老奔奔不但没有用蜡把它们封起来，反而还帮着它们往外钻。见到女儿们，她应该很高兴。她们弱不禁风，样子很怪，身上的毛湿漉漉地贴在身上，跟刚出壳的小鸡一样。她们的腿很软，摇摇摆摆的，几乎无法走到蜜杯那儿去吃早餐。吃完早餐，她们就爬回来蜷缩到妈妈身下，直待到身上的毛干了，变得毛茸茸的，身上也有劲了。

Only a few days more, and Old Bumble's daughters were ready to help her. Good, faithful,

cheery Old Bumble, her pretty wings had grown tattered and torn with her flying for food. Her fur coat, that was so fresh and fine, was now looking ragged. But the little daughters that snuggled up close to her soon began to do her flying, and she could stay at home and keep house and rest her tired wings.

才过了几天，老奔奔的女儿们就可以帮她干活儿了。善良、慈爱、快乐的老奔奔，由于长时间飞来飞去找吃的，那原本漂漂亮亮的翅膀现在已经变得破破烂烂的了。曾经那么鲜亮精美的绒外套，现在看着也褴褛不堪了。可是，依偎在身边的小女儿们不久就开始替她飞来飞去了。她可以待在家里管管家务，休整休整疲惫不堪的翅膀。

What was this stir and bustle about the nursery? Why, Hum and Buzz were ready to start out on their first journey. And little Flyaway was going, too. They looked much as their mother did when she left her bedroom, only much, much smaller. As their sisters were not quite ready yet to fly, they stayed at home and helped Hum and Buzz and Flyaway, when they came back, to feed the baby sisters that hatched from the eggs Old Bumble laid while they were spinning

Hum and Buzz were ready to start out on their journey and Flyawaw was going too.
哼哼和吱吱准备踏上他们的第一次旅程了。小翅翅也要去。

their cocoons.

育儿室里乱哄哄的，怎么回事？哦，原来是哼哼和吱吱准备踏上它们的第一次旅程了。小翅翅也要去。她们的模样和当初妈妈离家时一样，只不过小很多很多。妹妹们还不能飞，所以就留在家里，等哼哼、吱吱和小翅翅回来之后，妹妹们会帮着她们给新孵出来的妹宝宝们喂饭。她们给自己织茧衣的时候，老奔奔又产了一些卵，这些妹宝宝就是从这些卵里孵化出来的。

Now that the mother bee did not have to fly out of doors for the food, she laid eggs and eggs and still more eggs, so that the nursery was full of baby sisters of all sizes, from the egg to the cocoon age. As soon as the older ones came out of their cocoons, and had eaten and grown strong, they helped care for the growing babies. If they did not fly off to bring home fresh food, they stayed with their mother to help her with the nursery work.

现在，蜜蜂妈妈再也不用飞出去找食了，所以，她就不停地产卵，产呀产，产了一些又一些，结果，育儿室里挤满了大大小小的婴儿妹妹。有刚从卵里出来的，也有能够织茧作衣的。年长些的一从茧里出来、吃过东西、有了力气就开始帮忙照顾还没长大的宝宝。如果她们不飞出去往家里带新鲜食物,那就待在家里帮妈妈照顾小妹妹。

So the happy, busy Old Bumble and the daughters spent the days, doing together all the hard though pleasant tasks of bringing up a large family of children. Each day seemed much like another, and each new sister that crept out of her cocoon looked like her oldest sisters, only some were larger. Not one was as big as Old Bumble, though. Not one!

就这样,快乐、忙碌的老奔奔和女儿们齐心协力抚养着一大家子孩子,他们非常辛苦,但也非常愉快。每一天都很相似,每一个破茧而出的小妹妹也都和大姐姐们的模样差不多,只是有些个头要大一些。不过,谁的个头都比不上老奔奔。谁都比不上!

Much of their work was flying from flower to flower for nectar to make honey out of, and to gather pollen to take home in the baskets on their legs. And while they were doing this for themselves and their younger sisters, they carried pollen from flower to

flower, which helped make the seeds grow. So we can have the big red clovers and many other flowers they visit, so long as we do not destroy the bumblebees.

她们的大部分工作是在花丛中飞来飞去吸吮花酿蜜，把花粉收集起来放在腿上的篮子里带回家。虽然她们做这些都是为了自己和妹妹们，但是，她们把花粉从这朵花带给那朵花，也为种子的生长发芽助了一臂之力。所以，只要我们不毁灭大黄蜂，就总可以看到大红苜蓿花和许多别的大黄蜂经常光顾的花儿。

Time passed in this way until about the first of August. Then something different came to pass in this wonderful home. Some sisters crept out of their cocoons, who were as big as their mother—every bit as big as Old Bumble! And what is more, about the same time, some brother bees crept out of their cocoons: the very first brothers that Buzz and Hum and Flyaway ever had—Old Bumble's very first sons!

时间就这样一天天过去了，转眼就到了八月初。这时，这个奇妙的家庭变得不同寻常起来。一些妹妹从茧衣里爬出来，个头和她们的妈妈一样大——浑身上下每个部位都和老奔奔一样大！更奇怪的是，几乎同时，一些蜜蜂弟弟从茧衣里爬了出来，这可是吱吱、哼哼和小翅翅的大弟弟——老奔奔的大儿子呀！

These brothers did nothing to help their sisters who had tended them while they were growing up. These sons of the family did nothing to help their mother. But the sisters and their mother could do for themselves and each other, so it made little difference to them what the gay bumblebee lads did if they kept out of the way. And they did this for the most part, having a fine time of it, flying from flower to flower, eating as much as they liked, but taking nothing home for the others.

姐姐们照顾这些弟弟们长大，可是，这些弟弟们却什么活儿都不帮姐姐们做。这家的儿子们什么都不帮妈妈做。但是妈妈和姐姐们能自己照顾自己，还能相互照顾，所以，只要这些衣着华丽的黄蜂小伙子们不碍手碍脚，对她们来说，他们干什么都无关紧要。其实，大部分时间他们都在玩，在花丛里飞来飞去，想吃多少就吃多少，可就是什么也不往家里带。

And those August sisters, those big ones who looked like their mother—what did they, who were strongest of all, do in the home where they had grown up? Well, they sipped at the honey-mugs their smaller, older sisters had filled; and then, when they were strong, they sipped again. Some went to flowers, too; but some went only to the mugs in the nursery. They drank long, until their honey-sacks were filled. They needed food to last a long, long time, as a camel needs water enough to last him during a whole trip across the desert. It was their good-bye feast, and when it was done they went out through the doorway of that wonderful nursery. And they never came back again.

那么，那些八月出生的妹妹们，那些个头和妈妈差不多大的、最强壮的妹妹们，她们在生她养她的家里都做些什么呢？唉，她们吸着比她们个头小的姐姐们装在蜜杯里的蜜；然后，等有力气了，又接着吸。她们也有去采花蜜的，但有的只去育儿室吃蜜杯里的蜜。她们要喝很长时间，一直喝到蜜袋满了为止。她们要有食物来维持很长、很长时间，就像骆驼要有足够的水才能穿越沙漠一样。这是她们的告别宴。宴会结束，她们就走出门，离开那个奇妙的育儿室，从此，一去不复返。

Never!

They flew about a little while, and then, when they were ready, they crept into their bedrooms, each one by herself, and fell into their August doze.

永不复返！

她们四处飞一会儿，然后，等一切准备就绪，就爬进自已的小卧室，独自开始八月的小睡。

Ah me, the lazy things! There was clover yet, and many a sweet flower that needed its pollen carried for it, that it might have seeds. The smaller sisters, not nearly so strong, were still cheerfully at work, though their wings were tattered and torn and their fine fur coats were ragged. And these so fresh and big, with whole wings and new coats, weren't they ashamed to sleep the summer days away?

啊，我的天，这些懒东西！还有一些苜蓿花儿和许多香甜的花儿，可能还有种子，还等着她们搬运花粉呢。那些妹妹们，虽然个头比她们小，身体也一点儿不比她们结实，可是，却仍然愉快地忙碌着。她们的翅膀已经破破烂烂，漂亮的绒外套也已经褴褛不堪。她们这些翅膀完整、衣服崭新的大个子新人，居然在睡梦中消磨掉大好的夏季时光，难道她们不害臊吗？

No, they were not ashamed. It was not for them to tear their wings and spoil their clothes. Not yet! They had nothing to do in the world but to rest and save their strength. That is what their mother did the year before.

是的，她们不害臊。她们可不需要撕破翅膀、弄坏衣服。时候还没到！在这个世界上，她们除了好好休息、保存体力外，没事可做。一年前她们的妈妈也是这样。

So we will not call them lazy any more even if they did sleep so long. They rested while the hot August sun shone over the earth, and kept it all so warm that they did not sleep soundly, but moved now and then, as if they were dozing before a fire and just taking naps.

所以，即使她们真的睡这么久，我们也不要再说她们懒了。八月，烈日当头的时候，她们就睡觉休息。可是，地面热烘烘的，她们睡得并不踏实。她们时不时地扭动身体，就好像是在火堆旁边小睡，只不过是打盹儿而已。

They rested while the red and yellow leaves fell to the ground like a gay and beautiful shower! They rested while the snow followed after the leaves, and the cold made them numb in their beds.

当红的、黄的树叶像美丽欢快的雨点纷纷坠落的时候，她们在睡觉！当白色的雪花踏着树叶的脚步飘然而至的时候，她们在睡觉！天寒地冻中，她们躺在床上失去了知觉。

It was time enough to be up and doing when Spring called again. And shall we hope that when they wakened, brave and beautiful as their mother had been the year before — shall we hope that, when they flew low, humming their happy hunting song, the children of men had left spring flowers enough for the children of Old Bumble?

春姑娘再次敲门的时候，她们就该起床干活儿了。我们能否希望，当她们苏醒后变得像一年前的妈妈那样又美丽又勇敢的时候，当她们哼着欢快的采蜜曲低低地飞来飞去的时候，人类的子孙已经为老黄蜂奔奔的子孙留足了报春的花儿？

3

The Strange House of Cecid Cido Domy
锯蜂塞西德·西德·多米的怪房子

Cecid Cido Domy may not have been a witch, though she had a name that sounded as if she might have been one; and if she did not bewitch that willow branch, just what did she do to it? For, to begin with, the branch was just like any of the others on the willow; and by the time Cecid Cido Domy got through with it, it changed into a cone, shaped much like those that grow on Christmas trees.

也许塞西德·西德·多米不是女巫，不过她的名字听上去却挺像女巫。如果她没给那条柳枝施魔法，那她到底对它做了什么？这根柳枝原本跟其他所有的柳枝一模一样；不过塞西德·西德·多米折腾了它一通后，它就

The strange house of Cecid Cido Domy.
塞西德·西德·多米奇怪的房子。

变成了一个锥形小球，形状就和圣诞树上的那种球果差不多。

In fact, it changed before she got through with it, and there she was now right in the very middle of the very middle room. For the cone was a house with many rooms in it, though she herself never went outside the middle one—that is, she never did until she left Willow House for good.

其实，这根柳枝在塞西德·西德·多米还没折腾完之前就已经变成了小球果。此时此刻，她就躲在最中间那个屋子里的最中间。原来，这个球果是一所房子，里面有好多屋子，不过，塞西德·西德·多米从来没有离开过中间那个屋子，她会一直待在那里，直到永远离开这座柳舍。

There were strange doings in Willow House, too—much to interest anyone, with guests who came by day and by night and visitors who lived there just as if it was their house instead of Cecid Cido Domy's.

柳舍里有很多稀奇古怪的事情，多得大家已经见怪不怪了。客人不分白天黑夜地来串门，有的客人干脆住下来，就好像这房子是它们的，不是塞西德·西德·多米的。

One night a tiny creature came and found Willow House in the dark. She was dressed in soft gray, and when she was not flying, her wings folded down about her something like a long cape. The inner ones were gray as ashes, and the outer ones were darker gray, with bright brown upon them that was almost red; and they all had gray fringe that was softer and prettier than any other trimming could have been. She did not even knock, but she left an egg in one of the rooms, and her daughter, when she hatched out, lived right there, eating all the food she wanted and having a happy time in the shelter someone else had made ready, with no trouble at all to herself.

一天晚上，一个小东西飞到这里，发现了夜幕下的柳舍。这个小东西穿着柔和的灰色衣服，落下来不飞的时候，翅膀折叠起来搭在身上，像是披了一件长斗篷。小东西的翅膀里面那层是白灰色的，外面那层是深一些的灰色，上面点缀的褐色鲜艳得几乎成了红色。翅膀上镶着灰色的边，软软的，非常好看，比世界上所有的装饰边都软、都好看。这位不速之客连门都没敲就进来了,不过,她在一间屋子里留下了一枚卵。她的女儿出生后就住在这里,大吃特吃,开心极了。这可是别人造的房子呀!她根本没操过一点儿心,费过一点儿力。

Once another caller came in the day-time and left her egg in another room. She had no colors on her wings, and you could see right through them. The hind ones were fastened to the front ones by little hooks, so she looked as if she had only two wings when she really had four. She was a relative of a wasp, but she was not a wasp; and instead of a sting she carried a little saw. That is why people called her Sawfly. Her saw was a sort of tool she used when she was laying her egg, so that she could get it in far enough where it would be safe. Well, her daughter hatched out and made herself at home and ate all she wanted, just as the daughter of the gray moth had done.

有一次,另一位客人白天造访,在另外一间屋子里产了卵。她的翅膀没有颜色,是透明的。后翅膀和前翅膀用小钩子紧紧地钩在一起,所以,尽管她有四个翅膀,但看起来却像是只有两个。她是黄蜂的亲戚,但她不是黄蜂,因为,她随身携带的不是毒刺,而是一个小锯子。所以，人们又叫她“锯蜂”。她的小锯子是产卵时候用的，小锯子可以把产下来的卵送到最靠里面的安全地带。好了，她的女儿出生了，和灰蛾的女儿一样，她把这里当成了自己的家，想吃什么就吃什么。

Neither of these was any relation to Cecid, but the next guest was a sort of cousin. She made herself very much at home indeed, and left more than one egg at Willow House. She wasn't half as big as a mosquito, so of course her eggs were not large enough to take up much room. Her children, when they hatched, looked like Cecid, only they were even tinier; but they never went into the middle room, and as she did not go out, these cousins grew up in the same house without ever meeting each other.

这两位客人和塞西德扯不上任何关系，不过下一位客人却是她的一个什么表亲。她真的是一点儿也不客气，在柳舍里产了不止一颗卵。好在她的身子还没有半只蚊子大，所以，她的卵也不大，占不了多少地方。她的孩子们出生后，样子像塞西德，就是比她还要小。他们从来不进中间的那个屋子，而塞西德也从来不出那个屋子，所以，虽然在同一个屋檐下长大，但是，这些表亲们却从来没有见过面。

There were other guests, too, though none of them were invited to come. But it was all the same to Cecid Cido Domy. Guests might come and guests might stay or go! So long as they did not enter her middle room, or eat the food she needed for herself, or tear the house down too much, she was neither sorry nor glad. Indeed, she did not even know what went on in this wonderful house of hers.

还有其他客人，都是不请自来的。不过，对塞西德·西德·多米来说都一样。客人们想来就来，想住就住，想走就走！只要他们不进她中间的那个屋子，不吃她的那份食物，不把房子折腾得太厉害，她都无所谓，既不高兴，也不生气。真的，她甚至对她这奇妙的房子里所发生的一切一无所知。

And her house was pretty, too, as well as wonderful. It was shaped, as you know, very much like the cones that grow on Christmas trees; but the color was different. The shingles were soft and fuzzy and curved up about it like the petals of a rosebud, so that

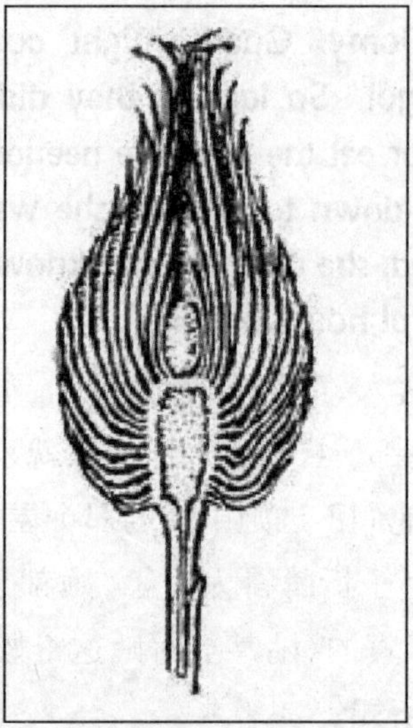

How Cecid' s house would look if it were cut in two.
如果把塞西德的房子一切两开就是这个样子的。

they were up-and-down and not the way houses are usually shingled. But you must remember that this was the house of Cecid Cido Domy, and not be surprised at anything about it. Or perhaps everything about it will surprise you, for it was all so queer and different from other houses.

她的房子不光奇妙，还很漂亮。你知道的，她的房子形状像圣诞树上结的球果，就是颜色不一样。房子的墙板软软的、毛茸茸的，像蔷薇花蕾的花瓣一样朝上曲伸箍在房子周围，所以，房子的墙面高低不平，一般房子的墙板可不是这样铺的。不过，别忘了这可是塞西德·西德·多米的房子，所以，别觉得奇怪。也许，这个房子的一切都会让你觉得奇怪，因为它真的太奇特，太与众不同了。

It may have looked as much like a rosebud as it did like a cone; for the shingles were stained creamy white at the small end, and were touched with pink across the middle, while the wide-curved outside end was silvery green. There never were prettier shingles made; and to think that, if it had not been for Cecid Cido Domy, each of them would have been just a

green willow leaf!

她的房子看起来像个球果，可是也像是朵蔷薇花蕾，因为，窄窄的墙板顶端呈乳白色，中间是一丝粉红，而宽宽的弧形边沿则是银绿色的。没有比这个更好看的墙板了。想想吧，如果不是塞西德·西德·多米，用来当墙板的就只有柳树叶了。

The way she changed green leaves into fuzzy shingles is a secret. Some people think that she poured out a juice on the tip of the twig in the spring, when the leaf-buds were ready to grow very fast, and that this juice had power to change the way they grew. All we really know about it for sure is that, whenever Cecid Cido Domy lives on a willow branch, a wonderful house grows up around her, and that the house is made of what started to be just green willow leaves.

没人知道她是怎样把绿树叶变成绒墙板的。有些人说，春天树叶刚刚发芽，正想快快长大的时候，她把一种果汁浇在了树梢上，果汁的魔力让它们变了模样。现在，我们唯一能肯定的是，只要塞西德·西德·多米驻扎到柳枝上，她周围就会长出一个奇妙的房子，而且，建这个房子用的是刚刚发芽的嫩柳叶。

The middle room had bare wooden walls, with the tiny brown boards going straight up and down. They met at the top in a point, but were not fastened together. That was the door to the room, and as Cecid stayed in and the other wee people in the house stayed out, this door was not used often. In fact, it was used but once, and that was the time Cecid Cido Domy went out never to come back again.

中间那个屋子的木墙光秃秃的，从上到下只有棕色的小木板。这些木板在屋顶聚拢在一起，但没有被钉死。所以，这个地方就成了屋门。塞西德总是待在屋里，而房子里的其他小客人们都待在屋外，所以，这个门并不常用。实际上，这个门只用一次，到时塞西德·西德·多米一出门，就再也不回来了。

But she had not yet left home, for the March winds were blowing and it was not warm enough. The pussies on the next willow bush had crept out of their little brown winter covers and looked as if they were sleeping in the sun. Some little birds with black caps on

their heads were flying quickly here and there, and were so happy that they sang "Chick-dee-dee, dee-dee," and then called " Fee-bee" sweetly, in a different tone, as if one song were not enough to show how joyous they were.

可她现在还没离家，因为三月还在刮风，天气还不够暖和。隔壁柳树丛里的银柳从褐色的小冬被里钻出来，好像是在闭着眼睛晒太阳。有些小鸟头戴黑色的帽子，疾速地飞来飞去，高兴得唱起了“吱—嘀—嘀，嘀—嘀”，然后换个调子，“啡—吡，啡—吡”地叫着，声音悦耳动听，似乎一支歌还不足以表达它们快乐的心情。

All this meant nothing at all to Cecid Cido Domy, who slept in her wooden room, as she had done ever since she went to sleep the fall before when cold weather came on.

所有这一切对塞西德·西德·多米来说毫无意义，因为她还在小木屋里沉睡。去年秋天寒气降临之前，她就进入梦乡了，一直到现在还没醒。

She was not much to look at when she went to

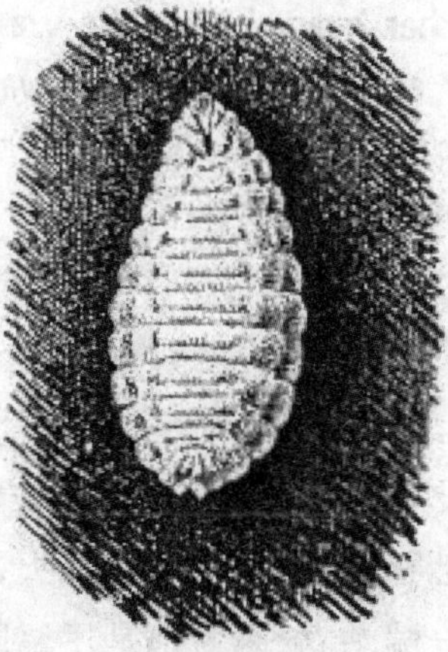

She was not much to look at when she went to sleep.
她睡觉时的样子没什么好看的。

sleep, only a little wiggly thing so short that it would have taken eight of her to be an inch long. She had no feet, and from her shape it was hard to tell which end her mouth was on. She had nothing that looked like a head, and one end seemed about as much like a tail as the other. The easiest way to tell which was which, was to see what end she stood on; for she stood on her tail and not on her head. Either she was yellow at one time and pink at another, or she was a color that looked different to different people; for some called her yellow and some called her pink.

她睡觉时的样子没什么好看的，只不过是个弯曲的小不点儿。特别短，八个加起来也才一英寸（2.54 厘米）。她没有脚，而且，从她的形状上看不出她的嘴在哪一端。她身上没有像头的地方，身体两端看起来都像是尾巴。要区分她哪边是尾巴，哪边是头，最容易的方法就是看她靠哪一端站立，因为她是用尾巴站立的，不是用头。她的颜色要么就是有时候黄，有时候粉，要么就是因人而异，因为有些人说她黄，有些人又说她粉。

No, she was not much to look at, being so small and having so little shape. But you can't always tell, from just looks, how much any living thing can do;

and Cecid Cido Domy was wonderful if she didn't look it—wonderful enough to get a home for herself and others, too, by making a willow-witch blossom into a rose.

的确,她这么小,也没什么体形,实在没什么好看的。不过,任何生物都不可貌相。塞西德·西德·多米的魔法让柳枝巫婆变成了玫瑰花，让自己和他人容身其中。如果这还算不上了不起的话,她还有别的了不起的本事呢!

She did another wonderful thing, too, before she went to sleep in the fall. That was after she had taken the last meal she was ever to have in Willow House, and after she had changed her dress a few times, and was as long as one eighth of an inch. She was old to be so tiny, for she had hatched out in the spring, and it had taken her until fall to grow even as big as that, though her house had grown up about her quite quickly.

去年秋天,她在入睡之前还做了另外一件了不起的事情。那时候，她已经吃完了在柳舍里的最后一顿饭,已经换了好几次衣服。那时候,她身长八分之一英寸。

她这小身材，和她的年龄极不相符。要知道，她是在春天孵化的，一直到秋天才长那么大，连她住的房子长的都比她快。

The wonderful thing she did before she went to sleep for the winter was to spin a little box all about herself. It was not a warm one, for it was so thin it could be torn by the slightest touch. But there was nothing in the middle room to tear it except Cecid, and you may be sure she didn't, after working so busily to make it.

冬眠之前她做的那件了不起的事情就是织了一个小盒子，把自己围了起来。小盒子并不暖和，因为太薄了，轻轻一碰就会破。不过，除了塞西德之外，中间屋子里没别人，所以，你尽管放心，她不会碰破这个小盒子的，这可是她辛勤劳动的成果。

Although she did not have enough of a mouth to make her head look like a head, it was as much as she needed, and it had a hole opening into it, out of which she could spill a sort of silk glue—the thinnest, finest glue you can imagine. As soon as it touched the air it

grew stiff so that it did not stick to her. In this way she made a little thin box all about her, somewhat as a caterpillar spins a silk cocoon. It was the color of water, so that she could be seen inside the box, as if it had been glass.

尽管她连一张像样的嘴巴都没有，头看上去也不像个头，但是她很知足。她头上有一个孔，从这个孔里她能吐出一种丝质胶，你能想象的最薄最好的胶。这种胶一接触空气就会变硬，所以不会粘在她身上。就是用这种方法，她做了一个小小的薄盒子，把自己围了起来，有点像毛毛虫吐丝做茧。这种胶无色透明，所以，从外面可以看到里面，就像是玻璃盒子一样。

There Cecid Cido Domy slept, standing on her tail, in her glassy silk box in the middle room of Willow House.

塞西德·西德·多米睡着了，在柳舍中间那间屋子里，在她那透明的、用丝织成的盒子里，尾巴着地站着睡着了。

While she slept a change came to her as strange as

the change which came to the willow when the tip of one of its branches turned into a rosebud cone. She changed from a maggot, with both ends pointed and looking much alike, into a tiny pink pupa, with something on her sides that looked the way wings do when they begin to grow.

她睡觉的时候，身体发生了奇怪的变化，当初那个柳树梢变成蔷薇花蕾小圆锥时也是这么奇怪。她变了，从一个两端尖尖的、一模一样的蛆虫变成了一个粉红色的小蛹，身体两侧看上去好像是要长翅膀似的。

Then she changed from a pupa into a little midget that looked something like a small-sized mosquito. By this time she was awake, and had pushed her way out of her glassy box and up, up through the middle of the middle room to the doorway where, for the very first time in her life, she stood where it was light. She was no longer in the dark, and she was no longer blind. Eyes had come to her in her great change, and she now had a good little head to wear them on and two long feelers to wave out in front. She did not stand on her tail any more, and she had six slender legs to hold her up and walk on. But she did not need just to walk, for

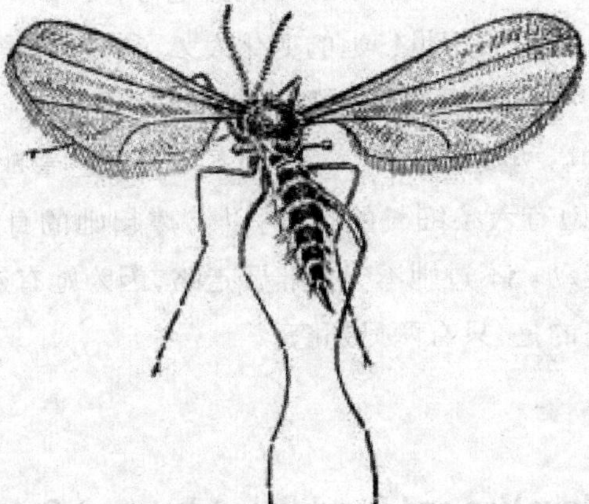

Then she changed into a little midget.
然后她变成了一只小型蚊子。

she had wings. There was something strange about them, though, for she had but two.

接着，她又变了。从一只小蛹变成了一个小不点儿，有点像一只小型的蚊子。这时，她已经醒了，已经从那个透明的盒子里挤了出来，挤过中间那个小屋来到屋门口，生平第一次站在了有亮光的地方。四周不再黑暗！不再什么都看不见！她的变化太大了。她有眼睛了，也有了一个像样的小脑袋，眼睛就安在小脑袋上。脑袋前面还伸着两根长长的触角。她不用再尾巴着地站着了，因为，她有六条细长的腿，可以支撑起她的身子，让她四处走动。不过她不一定非要走路，因为她有翅膀。不过奇怪的是，只有两只翅膀。

Now Van and Bumble and Jack and Carol all had four wings; and could Cecid Cido Domy get along with two?

Well, she had as many as a mosquito has, and two are enough for a bird or a bat! Besides, she had two pegs which helped her fly.

你可能会问，蝴蝶帆儿，黄蜂老奔奔，角蝉杰克和蚱蜢欢欢都有四只翅膀，塞西德·西德·多米只有两只翅膀行吗？

其实,她的翅膀和蚊子的翅膀一样多。对鸟儿和蝙蝠来说,两只翅膀就够用了!再说,她还有两只能帮她飞的平衡棒呢。

It was a warm day in spring when she stood by the door of her home that was now old with the storms of winter. It was about the time that willow-leaves begin to grow. So she could not stand there long, for she had an errand to do in the world. A very important errand it was, too. So important that, if she did not attend to it and at just the right time, there would be no willow houses for her children, and it would be a pity to have them hatch too late for that.

这个春日,她站在家门口,天气暖融融的。她的家经过寒冬风雪,已经破旧不堪。此时,正是柳叶生长的季节。她不能在那里站太长时间,因为,在这个世界上,她还有一件事情要处理,一件非常重要的事情。这件事情太重要了,如果她不能在恰当的时机把这件事情处理好,她的孩子们就没有柳舍可住了,为此而延误了孩子们的孵化,那就太可惜了。

Off she went—her two wings and two pegs were

enough, and it would never, never do to be too late! And here and there, near the bank of a little river, Cecid Cido Domy might have been seen that spring, seeking the willow-tips and laying her eggs one by one so that her little blind, legless babies might hatch when each willow-witch was just ready to have its leaves turned into shingles, stained creamy white and pink and silvery green, and curved like the pretty petals of a rosebud about the middle room, which should hide within its bare wooden walls a secret all its own.

她出发了。两只翅膀和两个平衡棒就够用了，而且，绝对、绝对不能太晚！那个春天，人们也许可以看到塞西德·西德·多米在小河的岸边，四处寻找柳树梢，一个一个地产卵，这样，她那没眼睛也没腿的小宝宝们就可以赶在柳树女巫刚要把柳叶变成墙板的时候孵化。那些墙板上撒满了乳白色、粉红色和银绿色的小斑点，就像蔷薇花蕾那漂亮的花瓣一样，向上弯曲着把小屋拢在中间，把属于她的秘密都掩藏在这些光秃秃的木板墙里面。

And so, when the March wind blows the fur of the pussies sunning themselves on the willow-twig, and

the happy bird with a black cap calls, "Chick-dee-dee, dee-dee" and, perhaps more sweetly still, "Fee-bee," is it very strange that we pick some twigs that have no pussies on them, but, instead, a little sleeping insect in a glassy silk box in the heart of a fuzzy cone—hoping that we may see, when the days are warmer, the little creature come forth and shake her wings? For fairy or witch or what-not, Cecid Cido Domy is a wonder well worth seeing; and if she does not bewitch the willow-branch, just what does she do to it?

当三月的微风吹动着柳树枝上晒着太阳的毛茸茸的银柳花时,当戴着黑色帽子的小鸟欢快地喊着“吱—嘀—嘀,嘀—嘀”,或者比那还甜美的“啡—吡”时,我们摘下一些嫩柳枝,上面不是毛茸茸的银柳花,而是一个毛茸茸的锥形小球,小球的正中是一个透明的丝盒子,盒子里面有一只酣睡的小虫。我们憧憬着天气再暖和一些的时候,这个小东西从里面爬出来展翅飞翔!这种感觉是不是很奇妙呀?无论她是什么,仙女也好,女巫也罢,塞西德·西德·多米的确是个奇迹,值得一看。如果她没有给柳枝施魔法,那她到底对它做了什么呢?

4

Poly, the Easter Butterfly
复活节的蝴蝶波丽

Easter came in April that year, which was much earlier than Poly, who lived in the north, would have come out of her chrysalis if she had not put it where she did. The fall before, when she was ready to spin a peg for her tail to hang on and a belt to hold herself up with, while she waited to be a butterfly, she left the parsnip plant she had been living on in the garden and took a very hurried trip. It was the first time she had ever been away from her parsnip home, but now that she had started, she seemed to be in a rush to get away from it. At the same time she seemed to be trying to get somewhere else as quickly as she could. So while

she was running away from one place and running into another, she climbed up the side of an empty house and crept in through a broken window. When she had gone as far as that, she settled down for the winter and turned into a chrysalis then and there—that is, she did so after she had spun a peg for her tail and a belt to hold her up.

那年的复活节是在四月，波丽住在北方，所以，假如她当初没把自己的蛹放在现在这个地方的话，那么复活节前她就从蛹里出来了。去年秋天，她匆匆忙忙地离开了花园，离开了她赖以为生的防风草。那个时候，为了能在等着变成蝴蝶的时候，有一个挂尾巴的丝栓子和一条支撑身体的丝腰带，她已经做好了吐丝的准备。那是她平生第一次离开她的防风草家园，可是，既然已经出发，她好像恨不得马上离开这里似的，又好像是恨不得马上就赶到另外一个地方去似的。就这样，在她从一个地方奔向另一个地方的时候，她爬上了一座空房子的墙，从一个破窗户钻进了房子里。到了这里，她才安顿下来准备过冬。也就是在那个时候、那个地方，她变成了一个蛹。当然，在变成蛹之前，她已经织好了挂尾巴的小栓子和拴身子的腰带。

Papil spun a peg and a belt and changed into a chrysalis just as Poly had done.
琶培儿也像波丽一样,织了一个尾栓和一根腰带,变成了一个蛹。

Soon after that, Papil, another caterpillar from another parsnip plant, traveled along much the same path that Poly had taken, and in much the same hurry to get away from somewhere and go somewhere else. He found the broken window, too, and went in and spun a tail-peg and a belt and changed into a chrysalis, just as Poly had done.

没过多久，另一只名叫琶培儿的毛毛虫也离开自己的防风草上路了。他沿着波丽的足迹，也像波丽一样迫不及待，恨不能立刻离开这里，奔向那里。他也发现了那个破窗户，也爬了进去，织了一个尾栓和一根腰带，变成了一个蛹。

The house that they chose would have been as good as an old shed or a fence-post if it had stayed empty. But it didn't. When the Soldier came back from France, his head was so tired that he wanted to go to the home where he had lived when he was a little boy, until he was rested.

他们挑选的空房子，如果一直是空着的话，那简直就能和破棚子或者篱笆桩相媲美了。但是，并非如此！士兵从法国回来了。他已经身心疲惫，恨不得马上回家，回到他儿时的家里，再也不离开。

That is how it happened that he found Poly and Papil when he was mending the window. At first he thought he would take them out to the shed. But when he touched Poly, she wiggled a little as if she wanted to be let alone. When he saw what a nice silk peg she had made for her tail to hold on to, and what a fine belt she had drawn about her to keep her in place, he didn't wish to take her down. So he let them both stay for company, and put flowering plants in the window so that they could have sweet cups to drink from when they came to be butterflies.

事情就这样发生了。修窗户的时候,他发现了波丽和琶培儿。一开始他想把她们拿到屋外的棚子里去。可是,他的手刚一碰到波丽,波丽就扭了扭身子,似乎不愿意让人动。看到她那个挂尾巴的小栓子那么漂亮,那条围在身上支撑着她的腰带那么精美,他不忍心动手了。于是,他留下他俩跟他做伴,还在窗台上放了些开花的植物,好等他们变成蝴蝶后,能喝到甜甜的花蜜。

Poly stayed in her chrysalis cell all winter, for it was not so warm by the window as nearer the fireplace. And then, one day in April, she pushed against her cell until the door opened and she came out

through the crack.

波丽整个冬天都待在蛹壳里，毕竟窗边不像壁炉附近那么暖和。四月，突然有一天，她开始使劲地往外挤，门开了，她从门缝里钻了出来。

She was a butterfly now, but her wings were not big enough to fly with. They were just four limp, floppy little flaps, hanging down her back. She had to cling with her six long slender black legs to keep from falling—Her wings were of no use. Four little wings good for nothing at all!

现在她是一只蝴蝶了，可是还不能飞。她的翅膀太小了，不过是四个小薄片，软塌塌地耷拉在背上。她用六条又细又长的黑腿拼命地抓住蛹壳，这才没有摔下来。没用的翅膀！四只小翅膀，一点儿用也没有！

But wait! As Poly clung to the thin wall of her winter cell, the wings were getting bigger! They were twice as big as at first and still growing! My, oh my, oh my! How they grew! Do you suppose they were getting big enough to carry Poly to the blossoms?

可是，别急！波丽拼命地抓着那薄薄的蛹壳的时候，她的翅膀一点儿一点儿地长大了！都是起初的两倍那么大了，并且还在长！天哪！哦，天哪！瞧它们长得多快！它们会长成大翅膀，把波丽带到花丛那儿去吗？

Yes, and it was Easter Day. There were tall white lilies and little blue lilies and other flowers, too; and they were so sweet that Poly spread her wings when they were strong enough, and fluttered over to the little blue lilies, and uncoiled her long black tongue and drank, for she was thirsty.

会的！那天是复活节。高大的白百合、小巧的蓝百合和其他花儿都已竞相开放。太香了！波丽的翅膀刚硬了，就迫不及待地飞到了小巧的蓝百合那里，伸出她那黑色的长舌头开始畅饮。她实在是渴了。

And there, drinking nectar from the next lily, was Papil, who had come out that very same day and had found the pretty sweet cups first. So in the large bay-window, full of flowers, Poly and Papil spent their Easter Day together, and, though they did not know it, the Soldier watched them a long while and smiled.

而此刻，在旁边的一朵百合花上，琶培儿也喝得正

When they become butterflies they drink nectar from flowers.
他们变成蝴蝶后便开始从花朵里吸食甘露。

香。原来那天他也出来了,第一个发现了这一杯杯香甜可口的花蜜。那宽阔的窗台上开满了鲜花,波丽和琶培儿就在鲜花丛中一起欢度起了他们的复活节。尽管他们并未察觉,可是,士兵却久久地注视着他们,脸上露出了微笑。

Butterflies among the flowers are beautiful on any day, but at Easter time they have a special meaning.

There is a reason why people paint downy chickens on their Easter cards and give their friends lilies at that time. The live chick comes from the still egg; the live lily grows from the dull brown bulb. Such things make us think how wonderful life is.

花丛中的蝴蝶每天都是美丽的。可是,复活节期间,她们还有特殊的意义。

复活节的时候,人们为什么要在贺卡上画毛茸茸的小鸡呢?为什么要给朋友送百合花呢?这是因为,活蹦乱跳的小鸡从静悄悄的鸡蛋中孵化而来;鲜艳的百合花从不起眼的棕色球茎中脱颖而出,这一切都让我们感受到生命的美妙。

So does the butterfly, who breaks the walls of the quiet chrysalis and comes out so full of life and beauty. Many years ago people loved the butterfly because of this, and gave her a sweet old name, "Psyche," which means "the soul."

蝴蝶也一样。她从沉寂的蛹里破壳而出，来到这个世界上，如此鲜活！如此美丽！许多年前，人们就因此而非常喜爱蝴蝶，还给她起了一个古老而又好听的名字"普赛克"，意思是"灵魂"。

It is no wonder that the Soldier smiled, with both lilies and butterflies to make life look beautiful to him that Easter Day.

Poly and Papil lived among the flowers for several weeks before Poly seemed to notice that there was a big pot of carrots and one of parsnips and one of celery growing there in the bay-window. These plants had no flowers, and why should Poly care about just leaves? She could not eat them if she tried, with her long tongue.

怪不得士兵会笑呢！那个复活节，百合花和蝴蝶交相辉映，让他看到了生命的美丽！

波丽和琶培儿在花丛中待了好几个礼拜。后来，有一天，波丽似乎才注意到窗台上还种着一大盆胡萝卜、一盆防风草和一盆芹菜。这些植物都不开花，只不过是些叶子，波丽凭什么要关注它们呢？就算她伸出长长的舌头去吃，也吃不下呀。

Ah, but there was a time when Poly had no long tongue—no tongue at all, in fact, for it was when she was only an egg; and in those days she had lived on the tip of a parsnip leaf. So we must not be surprised to find her, when she had eggs of her own, putting them on such leaves as she herself could have eaten when she was a caterpillar just hatched from an egg.

是呀！可是，曾经有段时间，波丽实际上根本就没有舌头，更别说长舌头了。那个时候，她还只不过是一个卵，终日寄居在防风草的叶尖上。所以，当她产下了自己的卵，并把它们放在这些叶子上的时候，我们千万不要感到奇怪。因为，她刚从卵中孵化出来，变成毛毛虫的时候，就吃这些叶子。

Now Poly in her caterpillar days could have eaten a great many kinds of leaves besides parsnips, if her

mother had left her on them; but I have never heard that she could eat anything that did not belong to the parsnip family. So if you find a caterpillar like Poly eating any sort of leaf, you may nod your head wisely and say, "That belongs to the parsnip family," whether you ever saw that kind of plant before or not. You see that, if you know a Poly caterpillar, she can teach you a great deal about plants. She never makes a mistake. Why, she would even go so far as to starve to death if any one tried to feed her a leaf that belonged to the wrong family. That is, when she is a caterpillar. When she is a butterfly, she drinks nectar from blossoms of many families and carries pollen for them all. But when egg-laying time comes, she flies back to the parsnip family just as if she remembered something.

假如波丽的妈妈当初没把她产在防风草上,而是产在了别的什么叶子上,那么,现在,变成毛毛虫后的这段日子，她可不就能吃到许多别的叶子了吗？但是,我可从来也没有听说过波丽能吃防风草科之外的植物。所以,如果你看到波丽那样的毛毛虫在吃什么植物的叶子,无论你以前见没见过这种植物,你都可以故作睿智地点点头说:“嗯,这是防风草科植物”。你瞧,如果你认识一个毛毛虫波丽，她就会教给你很多有关植物的知识。绝对不会错的！为什么？因为如果有人硬要给她喂

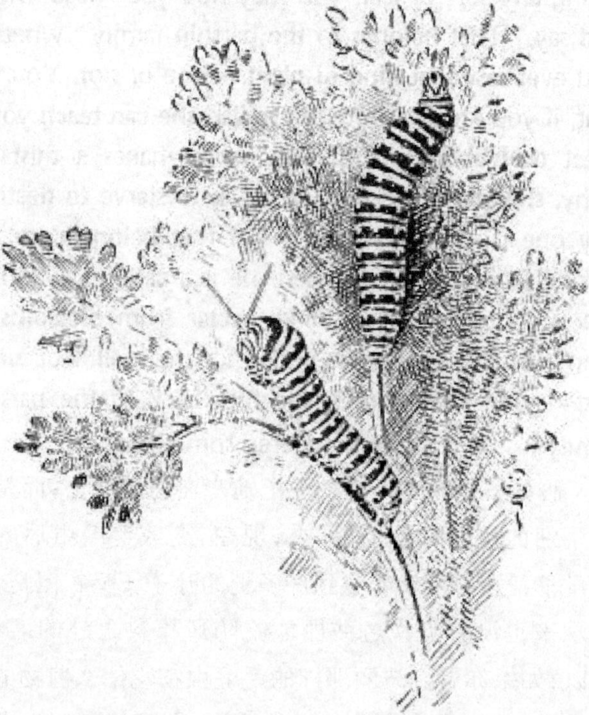

If you find a caterpillar like Poly, she can teach you a great deal.
如果你见到一只像波丽一样的毛毛虫,她可以教给你很多知识。

防风草科之外植物的叶子，她是宁愿饿死也不会张口的。当然,那是在她还是毛毛虫的时候。变成蝴蝶后,她会飞到其他科植物的花朵上饮用花蜜,还为它们搬运花粉。可是,产卵季节到来的时候,她总是像突然想起了什么似的,又飞回去找防风草科植物了。

Now some of the plants in this family are kinds that men, as well as Hexapods, like to eat, and we grow them for food. That is the reason we sometimes have to keep Poly's children out of our gardens, if there are so many of them that they eat more than we think we can spare. It is not hard to keep them out, for they do not hide and are easily seen. Some people let them have a few plants at the end of the row, because they like to watch the funny caterpillars, and because they love to see the beautiful butterflies they change into. That is what the Soldier did.

不光六脚节足小昆虫喜欢吃防风草科的植物,人类也喜欢吃。我们还把它们种来吃呢。所以,有时候,波丽的孩子太多了,余出的叶子不够她们吃,我们就不得不把她们赶出菜园。把她们赶出去并不难,因为她们不会躲藏,很容易看见。有些人会在地头留出几行防风草供她们享用,因为,他们喜欢看有趣的毛毛虫;还因为他们喜欢看她们变成的漂亮蝴蝶。那个士兵就是这样。

Zene, who hatched out of Poly's first egg, was contented enough when she crept from her shell on the tip of a celery-leaf. She hardly stopped to turn round before she began her green breakfast. Being a healthy baby, she was hungry, and ate most of the time except when she had to change her dress. This she did for herself from the very first, for Poly never did anything for her children after she had put her eggs on the right plants. She had taken care of her own dresses when she was a caterpillar, and if her daughter couldn't do the same, so much the worse for Zene!

从第一颗卵中孵化出来的是泽娜。那是在一片芹菜叶的叶尖上。她爬出来的时候,很知足。因为,她几乎连身都不用转,就可以享用那绿油油的早餐了。她是个健康的宝宝,所以很能吃。有时候她不得不换掉衣服,除此之外,大部分时间她都在吃。刚一出生,她就自己换衣服,因为波丽把卵安顿在合适的植物上之后,就再也不为自己的孩子做任何事情了。她自己是毛毛虫的时候,也得自己换衣服。要是她的女儿泽娜连这个也不会,那就糟透了!

But Zene could! Of course she could! Only it was hard to do, and it made her so nearly sick every time she changed her dress that for two days she couldn't eat a thing. She just spun a thin mat of silk and rested on that. You might think that Zene would have gone on wearing the same one if it was so much trouble to change.

不过泽娜会！她当然会！就是过程比较辛苦。每次换完衣服，她都大病一场，连续两天什么都吃不下。她只能勉强织一张薄薄的丝垫，躺在上面休息。你可能会想，要是换衣服这么麻烦，她完全可以一直穿着原来那套衣服不换呀！

Maybe she would, if she could. But she was such a hungry little thing that she grew larger than her dress, and her head got bigger than her skull, and grew out of it at the back, until it looked like a bunch inside the neck of her tight old dress. Of course, when she got as fat as that, there was nothing to do but push and jerk until she broke something. The first thing that ripped was a seam right around her collar, and of course when that tore—off dropped her skull!

也许，她可以不换，假如她能不换的话。不过，这个小东西太能吃了，所以，长的也太快，很快她的衣服就小得没法穿了。还有，她的头也越来越大，连后脑壳都快撑破了。后来，她的旧衣服紧紧地箍在越长越大的身子上，就像是一个扎紧了口的鼓鼓囊囊的小口袋。当然，胖到这个程度的时候，除了拼命挣扎、冲破束缚外，她别无选择。最先裂开的是领口四周的那条接缝，当然了，那条接缝完全撕裂之后，她的脑壳就掉下来了！

That gave her new head a chance to come out of the collar-hole, and then she crept right out after her head. And there was her old dress lying empty on the leaf.

Zene had a funny habit. Almost every time she changed her dress, she turned round after she had crept out of it and ate it up before she went on with her celery salad. It tasted good to her, so she did not waste it. When she didn't need it outside any more, she put it inside. She didn't eat her skull. That just rolled off the leaf.

脑壳一掉，她的新脑袋就趁机从领口洞里探了出来，接着，整个身子也爬了出来。而她的旧衣服则摊在叶子上，成了空壳。

泽娜有个有趣的习惯。每次换完衣服，她都会转过身子，把脱下来的旧衣服吃掉，然后才去享用她的芹菜沙拉。她觉得旧衣服味道不错，所以不想浪费。既然不需要穿在身上了，那就吃到肚子里。不过，她不吃脑壳。脑壳会直接从叶子上滚落下去。

Each dress lasted about a week before she had to change, and each new one was prettier than the one before. Her first ones had little bunches on them and a wide white sash shaped something like a saddle. They did well enough, but you ought to have seen her last dress. That was lovely! It was a pretty shade of green, with black bands for trimming, and every black band had a row of yellow spots on it.

每套衣服大约穿上一个礼拜就得换。所有的新衣服都比旧衣服漂亮。她最初的那几套衣服上都打着小小的褶，还有一条宽宽的鞍形白色饰带。那几套衣服的确不错，不过，你应该看看她最后换的那套衣服，简直太漂亮了！那是淡绿色的，镶着黑色的边，每条边上都点缀着一排黄色的小点。

It was a showy dress, Zene's last one. It could be

seen nearly as far as anyone could see the celery plant she lived on, and she never tried to hide. Even after the weather grew warm, and the Soldier put her out of doors, she didn't.

泽娜最后换的那套衣服光彩夺目。远远的，只要看到她那株芹菜，就能看到她。她从来不躲躲藏藏。就连天气变暖，士兵把她放到野外后，她也没有躲躲藏藏。

Now when you come to think about it, that seems queer. When Van was chased, she hid by making herself look like the bark on a tree. Carol sat still in the sand and didn't show. Jack had two ways of hiding, both good ones. Perhaps Ann Gusti's funny trick was the best of all, though Gryl did very well, as you would find out if you ever tried to catch him for a pet.

细想一下，似乎有点奇怪。蝴蝶帆儿被追逐的时候，她靠伪装成树皮藏了起来。蚱蜢欢欢则静静地坐在沙地上躲避视线。角蝉杰克有两个藏身的方法，都很有效。假死虫安妮有趣的藏身术也许是最高明的。不过，蟋蟀唧唧的方法也很不错，假如你曾经试图逮住他当宠物的话，你就知道了。

But Zene! There she stayed right on the upper side of the leaf and way out on the edge, no matter who came along.

Now, when an animal, whether it has four feet or more, does not hide itself when things come near it, there is some very good reason.

可是，泽娜呢！瞧，她就待在叶子上面最靠边的地方，不管谁过来都无所谓。

好了，当一个动物，不管它是四条腿的，还是四条腿以上的，在危险逼近的时候，躲都不躲，那么一定是有原因的。

Do you know what a wood-pussy is? That pretty animal, with black and white stripes and a tail like a plume, walks slowly and unafraid along the path. You can see him when he is still far off. If you wish to walk the other way, you have plenty of time. If you choose to go and meet him, he does not bother to turn out to the right to let you pass. He has shown you his black and white stripes as a warning. If after that you are foolish enough to come too near—you will wish you hadn't, that is all. The wood-pussy does not care. He doesn't mind the way he smells.

你知道臭鼬吗？那个漂亮的动物，身上是黑白相间的条纹，尾巴像个羽毛掸子。走在小径上时，总是慢悠悠的，一点儿都不害怕。从大老远你就能看到他。如果你想绕开走，完全来得及。如果你想走过去碰上他，也可以。可他不会闪到右边给你让路，他才懒得那么干呢。他身上的黑白条纹已经向你发出过警告了。假如你蠢到无视警告，依然执意前行，走得太近的话，那么你只有后悔的份了。臭鼬无所谓。他可不介意身上的气味有多难闻。

Now, can it be that Zene was unafraid during all her caterpillar days? Did nothing scare her? Did Daddy Bird try to take her home to his young ones? Was the white sash on her baby dresses to show where she was? Were the black and green stripes on her last pretty dress a warning?

那么，是不是在整个毛毛虫期间，泽娜都无所畏惧呢？真的没有什么东西让她害怕吗？鸟爸爸有没有试图把她带回家送给他的小宝宝们呢？她的宝宝衫上那白色的装饰带是在说“请勿靠近”吗？她最后换的那套漂亮衣服上的黑绿相间的条纹也是一种警告吗？

Well, something shook the leaf she was on one day, sharply and quickly, and this is what happened:

Zene threw up her head, and just behind it, from a slit in her back, she pushed out a pair of soft orange-colored horns. And just then there was a queer, oh, a very queer and very strong smell all around that celery plant.

谁知道呢？总之，有一天，她身下的叶子突然剧烈地摇晃起来。于是，发生了下面的这一切：

泽娜猛地抬起头，后背紧靠后脑勺的地方裂开了一个缝，从里面挤出来一对触角，橘黄色，软软的。也就是在那一刻，一种怪味，噢，非常怪、非常刺鼻的气味在芹菜四周弥漫开来。

I think it would have made you laugh if you had seen her. The Soldier laughed, and said something about a "gas-attack"; but that is soldier talk, and perhaps you will not understand what he meant.

看到她当时的样子，你一定会大笑。那个士兵就笑了，还说了一通什么“毒气战”之类的话。当然了，那是当兵的行话，你可能听不懂。

Zene's pair of horns was something like a pocket with two long pointed corners turned inside-out. In a moment she drew it down behind her head so that it was right-side-out, inside her body, and all out of sight; and the strange smell was soon gone.

泽娜的那一对触角，有点像是翻出了两个角的衣服口袋，那两个角又尖又长伸在口袋外面。眨眼之间，她又把触角收到脑袋后面，正面朝上放回体内，看不见了。那股奇怪的味道也很快消失了。

But though it could not be seen, the horn-pocket was there just the same, and already filled again with gas or whatever it was that Zene poured into the air when she threw her orange-colored pocket inside out and showed her horns. The stuff in her pocket is sour, and burns if it touches the tongue. Maybe Daddy Bird would not like that. So maybe Zene's bright stripes were one warning, and maybe the smell was another.

可是，看不见不等于没有。触角口袋还在她的身子里，这次，又装满了毒气。嗨，管它是什么气呢，反正在泽娜突然翻出橘黄色的口袋，露出那对触角的时候，那种气体就跑到空气里了。口袋里的东西已经发馊，而且烫舌。所以，可能是鸟爸爸不喜欢这个；可能泽娜身上那醒目的彩条就是一种警告；有可能，这股气味也是一种警告。

Perhaps her pretty dress, that could be seen so far, was like a danger-flag, and was a signal to other little creatures not to come too near her if they didn't like queer smells and sour, burny tastes.

她那从大老远就能看到的漂亮衣服或许也是一面警戒旗，警告其他的小生物，如果不喜欢闻怪味、不喜欢尝发馊烫舌的味道的话，就离她远点儿。

However that may be, it is certain that Zene never seemed scared, and never hid from anything until she was ready to throw away her last dress, pocket and all.

总之，不管是什么原因，有一点可以肯定。那就是，直到准备扔掉她最后那套衣服、包括口袋和其他一切的那一刻，泽娜才开始感到害怕，才开始躲躲藏藏。

When that time came, she left her celery plant in a hurry, and went somewhere else as fast as she could.

She climbed up the side of the house near where Poly had the fall before, but didn't go in through the window. That was mended. Instead, she prepared to hide on the outside of the house right in plain sight.

当那一刻到来的时候,她匆匆忙忙地离开了她的芹菜叶,拼命地奔向别处。她来到去年秋天波丽住过的那栋房子,爬上墙,波丽当时也是从这里爬上去的。但是她没有从窗户钻进屋里。那扇窗户已经修好了。她准备藏在房子外面众目睽睽之下。

First she rested for a whole day with her head pointed straight up and her tail pointed straight down. Once in a while she would swing a few loops of silk by moving her head, but not often. The second day she turned around so that her head was down and her tail was up. Then she went to work and spun silk from a tube that opened in her mouth. She swung out a silk thread as far as her head would reach to one side without moving the rest of her body. Then she grabbed the thread with her first pair of feet, which seemed more like arms, and pulled it back something as a man hauls up a rope "hand over hand." That made little folds in it, and she stuck the folded silk thread to

the board with silk glue. Then she swung out her head to the other side, and pulled back the thread with her "hands," and glued down that folded part. This was the way she made her peg of silk, swinging her head from side to side and pulling back each thread in folds and sticking it down. It was a firm peg and a little curved, and it took her about half an hour to make it.

她先头冲上、尾巴朝下休息了一整天。这期间，她偶尔会晃动脑袋吐出几个丝环，但是，只是偶尔。第二天，她转了个身，头冲下、尾巴朝上开始工作。她的嘴里有一个软管，开着口，她就从那个软管里往外吐丝。她使劲地伸头吐丝，头伸多远，织出来的丝线就多长，这期间，身子的其他部位却纹丝不动。接着，她用最前面的那双脚紧紧钩住吐出的丝线，像人两手交替拉绳子那样，把它拽回来。唉，她的那双脚看起来更像是胳膊。这样把丝线对折起来之后，她又吐出丝质的胶把对折好的丝线粘在木板上。然后，她的头又摆到另一边，用“双手”把丝线拽回来对折好，粘起来。就是这样，她的头从这边摆到那边，把丝线一根根地对折起来粘住。最后就做成了她的小丝栓。这个丝栓很结实，稍稍有点弯曲，这项工作花了她大概半个小时。

As soon as it was done, she turned round again so that she was head up and tail down once more. Her tail seemed to be in a hurry to find the peg, and felt around for it as if it were hunting for something. After lifting her tail up and feeling to one side and then the other, she at last hit the peg, and then she took right hold of it with the hooks in her last pair of feet and tangled them up in the silk threads and pulled tight until she held fast. She had hitched herself by her own tail and couldn't get away.

刚做完小丝栓,她就又转过身子,重新回到头冲上、尾巴冲下的姿势。她的尾巴大概是急于找到小丝栓,只见它来回扫着,好像是在搜寻什么似的。后来,泽娜翘起尾巴在这边扫扫,又在那边扫扫,总算是找到了小丝栓。于是,她立刻伸出最后面的那双脚钩住了小丝栓,她把脚上的那些小钩子伸到丝线里绕了又绕,拉了又拉,直到她能抓牢为止。她是用自己的尾巴把自己吊起来的,所以,脱不了身了。

But she was not yet ready to be a chrysalis. For Zene was not going to hang head down like Van while she waited to be a butterfly. No, she was going to sleep head up, so she needed more than a tail-peg to hold her.

可是,她还没做好变成蛹的准备呢。泽娜可不打算像帆儿那样头冲下,等着变成蝴蝶。不行!她睡觉的时候,要头朝上。所以,光一个尾巴栓可不行。

What she needed was a loose belt reaching from the board around her body. So she spun it. She glued the end of the first silk to the side of the house, and then spun a loop in front of her, stretching it up between her first and second pairs of legs, using the first pair like arms to catch it into place when she had spun it long enough so that her head could bend down and fasten the other end to the house. Thus she spun her belt, holding it up something like a skein of yarn in her "arms," and gluing the end to the board each time. Once it broke while it was still very weak, and she fell way over to one side. But brave patient little Zene, with her tail hitched fast, felt her way back into place and began all over again.

她还需要一根腰带,一根能从木板上吊下来,把她轻轻地围住的带子。于是,她又开始织带子。她先把第一根丝的一头粘到墙上,然后在身子前面绕成一个环,用最前面的那两双腿撑住,当丝线吐到足够长的时候,

while she was a caterpillar, and it certainly didn't look as Zene would when it came time for her to be a butterfly.

哇，那个长着一个滑稽的狮子鼻和两只可笑耳朵的小东西真的是泽娜吗？可它长的一点也不像毛毛虫泽娜呀！它的长相也绝对不像准蝴蝶泽娜呀！

But Zene was inside that queer-shaped little case, for all that. And at last she had hidden; for the case soon grew to be about the color of the board on the side of the old house, and she could not be seen far away.

可是，不管怎样，泽娜就在那个奇形怪状的小匣子里。终于，她藏起来了。因为，那个匣子的颜色很快就变得差不多和墙板一样，从远处看不到她了。

Zene did not sleep all winter like her mother in the chrysalis case, but changed into a butterfly in about two weeks, and laid her eggs on some caraway leaves.

藏在蛹匣子里的泽娜，并没有像她妈妈那样整个冬季都睡觉。大约两个星期后，她变成了一只蝴蝶，然后，在一些葛缕子叶上产了卵。

It was still summer when her children hatched, and Poly, the oldest and named for her grandmother, did all the things her grandmother had done except getting inside the Soldier's house to make her chrysalis. It was late in May when she wakened in the spring. But as butterflies have their own calendars instead of ours, perhaps it was her Easter Day, just the same.

孩子们孵化时，夏季还没结束。为了纪念她的祖母,第一个孩子也取名波丽。她祖母做过的事情,她都做了,只有一件事情除外:她没有钻进士兵的房子里去做蛹。春天,她醒来的时候已经是五月底了。可是,蝴蝶的日历和我们不一样，她们有她们自己的日历，所以，对她来说,那天就是复活节。

5

Jumping Jack
蹦蹦跳跳的角蝉杰克

You would never have thought, to see Jack, that he could jump. He looked as if he had grown on the bitter-sweet vine. He looked like a thorn that would stay there even in winter, like the bright red berries.

看到杰克,你绝对想不到他会跳。他看起来就像是长在苦甜藤上的一根刺,和鲜红的浆果一样,即使是冬天,也会待在树上不肯下来。

But when you stopped to think about it, you would remember that a bitter-sweet vine does not have

When you stop to think about it you remember that a bitter-sweet vine does not have thorns.
你只要定神想一想,就能想起来苦甜藤是没有刺的。

thorns—not real ones. Then you would point your finger at Jack and say, "You can't fool me, sir! You are not a thorn"; and before you got through telling him that, Jack would be gone.

不过，如果你静下心来想一想，就会记得苦甜藤上其实根本就没有刺，没有真刺。于是，你可能会指着杰克说"喂，先生！你蒙不了我！你不是刺！"可能你话还没说完，他就不见了。

He wouldn't jump, though. Not that time! He would just slip around to the other side of the branch, where you could not see him, and sit there looking as if he had not moved, and as if that were the place he had always grown in. So you would play hide-and-seek with him for a long time: Jack would hide and you would seek.

不过，他不是跳走的。那个时候他不跳！他只是溜到了树枝的另一侧而已，在那儿，你就看不到他了。他原模原样地坐在那儿，好像根本没有挪过地方，好像他一直就长在那里。所以，你会和他玩很长时间的"捉迷藏"：杰克藏，你来找。

But if Jack is touched too hard, then he is nimble and he is quick and he will jump off with a snap and spread his four wings and fly a little way.

可是,如果被逼急了,他也是身手不凡哟!他会啪的一声跳起来,张开四个翅膀,飞到不远处。

For Jack can fly, and so can Jill, his little mate; but they will both come back again to drink the sap of the bitter-sweet vine.

杰克会飞了,他的小伙伴吉尔自然也会了。不过,他俩都会再飞回来喝藤蔓里那又苦又甜的汁液。

There they sit now, with their sharp slender beaks stuck right down through the bark, and drawing the juice up through the hollow tube, as you suck lemonade through a straw. They find it pleasant to sit there all the summer day and sip the juice of the vine.

这会儿,他们就坐在藤蔓上,又尖又细的嘴巴扎进树皮里,吸着藤里的汁液,就和你用吸管喝柠檬汁一样。他们整个夏天都坐在那里,喝着藤蔓里的汁液,真惬意呀!

个卵兄卵弟、卵姐卵妹同住一室，而吉尔的兄弟姐妹比他的还多。

Jack was the last egg his mother put through the hole in the bark into the nest. When he was safely poked down at the end of the row, she covered the hole with something that looked like the nice sticky frosting that is dripped from a spoon to the top of a cake. It wasn't sweet, so it couldn't have been frosting; but that is the way it looked, and it was in a wavy heap almost as big as Jack's mother.

杰克是妈妈从树皮上的洞里放到窝里的最后一颗卵。妈妈把他稳稳当当地捅到他那一排的末尾放好后，就用一种细细的、黏糊糊的东西把洞口盖了起来。那种东西很像是从勺子上滴到蛋糕上的糖霜。它不甜，所以不可能是糖霜。可是，看上去就是，而且一圈一圈地堆成一堆，几乎和杰克的妈妈一样大。

Now wasn't that a good way of tucking her eggs in—to plaster them down with a sticky white blanket? Jill's mother made her nest just the same way. So there they were, snug as could be, little Jack and Jill and all

their brother-and sister-eggs.

你说,这种把卵塞进去、用一条黏糊糊的白毯子盖住的方法是不是很妙呀?吉尔的妈妈也是用这个方法做好了自己的窝。现在,你瞧他们,小杰克、小吉尔和他们所有的卵兄卵弟、卵姐卵妹们都躲在里面,可真是要多舒服有多舒服呀!

Think of the things that were to happen before they could hatch! Ann Gusti was still playing her clown tricks in the meadow, with the big blue sky for a tent, several weeks after Jack and Jill were tucked into their beds. Gryl sat before his open door, and fiddled a happy lullaby of "Cri-cri-cri." Luna, brown as a nut, lay on the ground in her own silk bedroom, fast asleep, too.

想想他们孵化之前都会发生些什么事吧!杰克和吉尔被塞到床上都好几个星期了,可是,假死虫安妮却还在草地上玩着她的小丑把戏,广阔无边的蓝天就是她的帐篷;蟋蟀唧唧则坐在敞开的门前,唧啾——唧地哼着快乐的催眠曲;全身像坚果一样棕褐色的月形天蚕蛾卢娜也躺在地上她自己的小丝屋里睡得正香。

Cold weather came, when the ice did not melt even in the daytime and when the chilly earth was wrapped in a deep blanket of snow, as white as the cover Jack's mother had put over him and much thicker.

天冷了。大白天都在结冰。寒冷的大地盖着一条厚厚的雪毯，白得就像杰克的妈妈盖在他身上的那条毯子，当然，比它厚多了。

There were a few days in January when the snow thawed, and Van wakened in the hollow tree, and came out, and flew about the sunny places in the lane, and drank at the edge of a little pool.

一月份有那么几天，雪融化了。树洞里的帆儿醒了。她爬出树洞，在小路上找有阳光的地方飞了一会儿，还在一个小水塘边上喝了点儿水。

Candlemas Day came, and Old Bumble was so sound asleep she never even buzzed.

圣烛节到了，黄蜂老奔奔睡得太香了，甚至连嗡都没嗡过一声。

And all this time Jack slept under his blanket, that was wrapped so closely to the twig that the snow could not sift in, and stuck down so tightly that the wind could not lift the corners.

Even when the willow pussies crept out along the twigs to warm their fur in the March sun—even then Jack was asleep.

这期间,杰克一直盖着毯子沉睡。毯子紧紧地裹着树枝,掖得严严实实,雪渗不进去,风也甭想掀开。

就连毛茸茸的银柳花悄悄地爬满枝条,在三月的阳光下晒太阳时,杰克都在睡觉。

Some pretty blue lilies blossomed for Easter Day, but Jack was still a tiny Easter egg himself, and it was not yet time for him to hatch.

为了迎接复活节的到来,许多漂亮的蓝百合也开放了。可是,杰克仍旧是一个小小的复活"蛋",现在,还没到他孵化的时候。(注:在西方,复活节人们把蛋染成各种鲜艳的颜色,把它作为礼物送给朋友,表达友谊、关爱和祝愿。复活蛋也象征着生命的开始与延续。这里,egg:卵,也有蛋的意思。这里是一语双关。)

And then at last May came and woke him. How did she call him when he had slept so long and so soundly? Perhaps with her sunshine, which brooded the nest of eggs in the bittersweet vine with its warmth, as a mother hen broods over her nest of eggs in the hay with her cozy feathers.

终于，五月姑娘姗姗而来叫醒了他。杰克睡了那么长时间、睡得又是那么香，她是怎么叫醒他的呢？也许是她那温暖的阳光孵化了苦甜藤里的这些卵，就像母鸡用暖融融的羽毛孵化干草窝里的鸡蛋一样。

And here he was—a feeble baby, so wee you would need a magnifying glass to see him with! What was he to do with that heavy blanket over him? If his mother had stuck it down so tightly that even the strong north wind could not move it, how was Jack to get out of bed? Poor little Jack with his twelve brothers and sisters! Whatever were they to do?

他孵出来了，多么虚弱的一个小宝宝呀！太小了，你得用放大镜才能看清他。他身上那条又厚又重的毯子可怎么办呢？她妈妈当初把毯子掖得那么紧，就连呼啸的北风都掀不开，那么，现在，杰克可怎么起床呀？好可怜呀，小杰克和他那十二个兄弟姐妹！他们到底该怎么办呀？

Well, here he comes, like a tiny Jack-in-a-box, poking his head right up through the white blanket. He does not try to lift it; he just sticks up his yellow head and red eyes, and pushes himself out of bed.

瞧，他出来了！直接穿透白色的毯子伸出了脑袋，就像是“精灵盒”里的小精灵。他没有掀毯子，只是挺起自己黄黄的脑袋和红红的眼睛，从床上钻了出来。

Of course Jack is thirsty, and of course a baby who can get out of his nest all by himself the first day can feed himself, too. So out he creeps to the tender leaf, and digs his little beak down into it, and takes his first drink of sap.

当然了，杰克很渴。同样理所当然，一个出生第一天就凭自己的本事出窝的宝宝，也肯定能自己养活自己。他爬到一片嫩叶跟前，把小嘴刺进去，喝了他有生以来的第一口树汁。

He likes it! Oh my! my! how he likes it! He likes it so well that for seven whole days he hardly stops drinking.

他喜欢！噢，天哪！天哪！太喜欢了！喜欢得几乎停不下嘴，一连喝了整整七天。

By that time his first baby clothes are so tight that he can not swallow another drop. What is the greedy youngster to do about that? His clothes are grown on him, too! Think of it!

那个时候，他的第一身宝宝服已经绷得太紧了，多一滴都喝不下去了。贪吃的小家伙这下该怎么办呢？他的衣服可是长在身上的呀！想想吧！

Never mind. Baby Jack is so strong now, after taking a drink that is seven days long, that all he has to do is to stretch himself; and this rips his clothes in a little seam near the back of his head and tears his baby cap. Now he can pull his head out of the old cap and then get out of his first baby clothes through the torn place.

别担心！连喝了七天的杰克宝宝现在已经很健壮

了，他只需伸伸懒腰，后脑勺那儿的一条衣缝就裂开了，婴儿帽也破了。现在，他能把头从旧帽子里伸出来，然后，从撕破的地方钻出来，甩掉他的第一套宝宝服了。

His first dress had long hairs on it, but his second one, which grew on him under the old one, does not have these hairs. Instead, it has six lumps in a row on the middle of his back. You would think that the second dress, growing under the first one, would be even smaller and tighter.

他的第一套衣服上有长毛。第二套衣服也长在身上，就在第一套衣服下面，却没有这些毛。不过，背中线上却鼓着排成一列的六个包。你可能会想，他的第二套衣服既然长在第一套衣服下面，那么，第二套衣服一定会比第一套衣服还要小还要紧。

But as it stretches as Jack grows, he can take another long drink before his second dress is stretched as big as it can be. So he puts down his beak again and drinks for six more days. Then he needs a third dress,

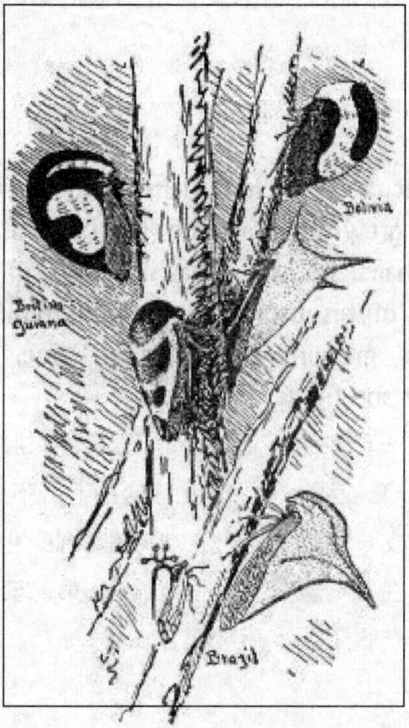

Jack' s foreign cousins
杰克的外来表亲们

which he gets in just the same way as before—he creeps out through a torn place at the back.

可是，杰克的第二套衣服却能随着他的成长而伸展，所以，在第二套衣服撑破之前，他还能再喝个够。于是，他又一次伸出嘴巴，连喝了六天。然后，他又要换第三套衣服了。和以前一样，这套衣服也是在他从后背撕破的地方钻出来之前就已经准备好了。

This time he still has the six lumps in a row in the middle of his back, and, besides, a hump has begun to grow at the back of his head. While he looks like that, he takes his third drink of bitter-sweet sap, and this lasts about six days, too.

这次，他的后背中线那儿依然鼓着排成一列的六个包，另外，他的后脑勺那儿也开始往外鼓包了。带着这副尊容，他第三次喝起了又苦又甜的树汁，这次也连喝了六天。

After that he gets his fourth suit of clothes—much like the third one, lumps and all; only the hump on his shoulders is bigger, and there are two little flaps

below on each side. These clothes keep on stretching for two whole weeks, so that Jack has a chance to drink for fourteen days without stopping for his next dress.

之后，他换上了他的第四套衣服——很像第三套，连鼓着的包都很像。只不过肩膀上的包大一些，而且身体两侧靠下面的地方多了两片小羽翼。这套衣服一直伸长了整整两个星期，所以，杰克趁机连喝了十四天，连下一套衣服都没顾上换。

But dear, dear! When he creeps out of his fourth dress, what a hump he does have! It has grown way up over his head. Yes, Jack is a hump-back, and all the doctors in the world can not cure him. And the two flaps on each side are much, much bigger.

可是，天哪，天哪！当他脱下第四套衣服的时候，你看他身上的那个包！已经高出头来一大截了。是的，杰克是个驼背。世界上所有的医生都治不好他。他身体两侧的那两片羽翼也比以前大了很多很多。

Well, never mind! The six lumps down the

middle of his back and the hump on his shoulders and the flaps on his sides do not bother him a bit. He is as healthy and as thirsty as ever. By this time he is strong enough to dig his beak right down into the stem of the vine, where the sap is running freely. All his clothes are pale at first, but grow darker as they get older, when they are gray or brown, and trimmed with red and sometimes with a little white.

不过,不用担心!他后背中线上的六个包、肩膀上的那个包，还有身体两侧的羽翼，一点儿都没让他烦恼过。他还一如既往地健康、好饮。这个时候,他已经非常强壮了，能把嘴直接插进藤茎里，去喝里面的汩汩汁液。他的衣服起初都是浅色的,后来,随着衣服变旧,颜色也就变深了,成了灰色或者是褐色,上面还掺杂着一些红色,有时候还会带一点儿白色。

Jack sits and sips and grows in his fifth dress just as he did in all the other four; and at last this one becomes too tight just as the others did, one after another. That does no harm, for he can still have one more suit of clothes. So when these get too tight, he stretches and stretches inside of them, until they rip open and he can crawl out.

杰克穿着他的第五套衣服坐在那儿，一边喝，一边长，就和他身着前四套衣服时一样。前面那几套衣服都是一套接一套紧得没法穿了，这套衣服也一样，最后也紧得没法穿了。不过，这不碍事，因为他还有一套衣服可换。所以，当这身衣服紧得没法穿的时候，他伸起了懒腰，是在衣服里面伸的，一直伸到衣服裂开，他能从里面爬出来。

When he is all out and has rested a bit, he puts his beak down into the bitter-sweet vine and goes on drinking as if nothing much had happened.

出来休息了一会儿之后，他又把嘴扎进了苦甜藤里，接着喝上了，好像没发生什么了不起的事似的。

Why Jack, don't you know that a wonderful change has come to you? Don't you know that the hump on your shoulders is so big that it makes you almost twice as tall as you would be without it? Don't you know that four of your legs now look something like tiny leaves? Don't you know that your two hind legs are longer than they were before, and that you can

jump? Oh, how nimble and quick you can be when you jump! And don't you know, you funny Jack, oh, don't you know that the flaps on your sides have changed into wings and that you can fly away and back again? Don't you know—don't you even know that you are no longer a growing baby but a grown-up treehopper?

No matter how much you may drink now, you cannot be any bigger, for this is the last suit of clothes you can ever have—the very, very last!

可是,杰克!你可知道,你已经今非昔比了?你可知道,你肩上的那个包已经长得和你一样高了?如果没有它,你现在可没这么高呀。你可知道,你的四条腿现在看上去有点像小树叶?你可知道,你的两条后腿比以前长了,你能跳起来了?哇!你跳起来的时候一定身手不凡!还有,可爱的杰克,你可知道,你身体两侧的羽翼已经变成了翅膀,现在你能飞出去又飞回来了?你可知道,你不会连这个也不知道吧,现在你已不再是一个小宝宝了,你已经长大变成角蝉了?

现在你喝得再多也不会长个儿了。因为,这是你的最后一套衣服了!的的确确就这一套了!

How much of all this do you suppose the queer little chap really understands, as he sits on the vine looking like a thorn? Well, just as much as Jill knows, who hatched on a May day, too, and grew up in the very same way, humped back and all.

But while Jack can only sit and sip and jump and fly and play hide-and-seek with you as the summer days go by, Jill finds something else to do before the weather gets cold.

这些话，你觉得这个坐在藤蔓上像个刺一样的怪家伙能听懂多少呢？嗯，他知道的应该和吉尔差不了多少。因为吉尔也是五月份孵化的，也是这样长大成蝉的，连驼背都一样。

可是，夏日一天天过去，杰克只会坐在那里，喝呀、跳呀、飞呀的，还和你捉迷藏，而这个时候，吉尔却趁着天还没有变冷，找了点别的事情做。

Jill has a neat little tool at the end of her body, and it would be a pity not to use it. For one thing, it is a cunning knife, and with it she cuts a hole in the twig and digs out a nest for her eggs. For another thing, it is a sort of bubble-blower, and with it she stirs up her foam that looks like frosting. And last, it is a little spoon

that she uses to drip the pretty white sticky stuff down over the nest of eggs. Slowly, back and forth and down under and up over, she moves her spoon, standing all the time on tiptoe to make the wavy rows of foam so very, very carefully.

吉尔的身体末端有一个精巧的小工具,要是不用就太可惜了。一来,它是一把精巧的小刀,有了它,她就可以割开嫩枝,挖一个小窝安顿她的卵。二来,它是一个泡泡机,有了它,她可以制作像糖霜一样的泡沫。此外,它还是一把小勺,有了它,她可以把那种白白的、黏糊糊的东西滴到卵巢上面去。她一直踮着脚尖,慢慢地、上下左右、来来回回地转动着她的小勺,小心翼翼、专心致志地堆出一排排像波浪一样的泡沫。

For this is the foam that stiffens into a blanket to cover her eggs; and as she can not stay on the cold twig to take care of them through the winter days that will be coming, she must do what she can to make the blanket just right.

就是这种泡沫,凝固后能变成一条毯子,把她的卵盖起来。冬天即将来临,那时她就无法留在冷飕飕的小树枝上照顾他们了,所以,她必须尽可能地把这个毯子

This is the foam that stiffens into a blanket to cover her eggs.
就是这种泡沫,凝固后能变成一条毯子,把她的卵盖起来。

弄得妥妥帖帖。

She makes a good one, little Jill does: one so like the one that her mother put over her and the one that Jack's mother put over him when they were eggs in a nest, themselves, that you could not tell which of the three was the best.

她毯子做得不错！小吉尔做的毯子的确不错！就像她的妈妈盖在她身上的那条毯子，也像杰克的妈妈盖在杰克身上的那条毯子。那个时候，他们自己还是待在窝里的卵呢。你真的无法知道这三条毯子中，哪一条最好。

So there they will stay, tucked in snug and safe. Jack Frost will be about when the cold nights come, but he will not harm them. North Wind will whistle through the bittersweet vine and shake the red berries, but he can not lift the blanket Jill stuck down. And many other strange and wonderful things will happen during the days of fall and winter and early spring.

所以，孩子们将会待在这里，暖暖和和、舒舒服服、安安全全地挤在一起。寒夜里，霜大哥会来做客，但是，他不会伤害他们。北风大叔也会从苦甜藤里呼啸而过，让火棘果瑟瑟发抖，可是，他却无法掀开吉尔盖得严严实实的毯子。当然，秋天、冬天和早春时节，还会有其他许多奇怪而又奇妙的事儿发生。

But, after all, nothing will be stranger and more wonderful than when May calls Jill's babies, and each little yellow head with bright red eyes comes popping up through the white fluff like a Jack-in-a-box.

A May day is a pleasant time, when things like that are happening out on the bitter-sweet vine!

然而，说到底，最奇怪也最奇妙的事儿莫过于吉尔的宝宝们从窝里出来时的情形了。听到五月姑娘的呼唤，他们就砰的一声穿透盖在身上的白绒毯，露出一个个黄黄的小脑袋，脑袋上红红的眼睛亮晶晶的，就好像“精灵盒”里的小精灵，一摁机关，就从里面弹了出来。

五月的日子愉快又惬意，因为苦甜藤那边，每天都在上演这样的故事！

6

Nata, the Nymph
蜻蜓娜塔，水中仙子

（注：Nymph：希腊罗马神话里居于山林水泽的女神；也指美丽的女子。有时候，人们把蜻蜓叫作 water Nymph，而蜻蜓的幼虫若虫的英文也是 nymph。）

Once upon a time Nata could not walk or fly or swim. Of course that was when she was an egg.

While she was a helpless little thing like that, her mother dropped her into the water and flew off and left her.

以前，娜塔不会走，不会飞，也不会游泳。当然，那时，她还是一颗卵。

当时，她还是那么一个小不点，不能自立，她的妈妈就把她丢进水里飞走了，离开她了。

This sounds cruel, and when you know that Nata's mother was the creature that people sometimes call the devil's darning-needle, and that children sometimes talk about, with scared looks on their faces, when they say that she will sew up their lips and maybe their eyes and ears, too, if she catches them—when you know that was Nata's mother, maybe you will think she was just wicked enough to let Nata drown.

听着有点残忍，是吧？如果你知道，娜塔的妈妈就是那个有时候被称作“魔鬼缝衣针”的家伙，就是那个孩子们一提到就会面露恐惧，说一旦被她抓住，嘴巴就会被缝起来，甚至连眼睛和耳朵也会被一起缝起来的家伙，如果你知道，她就是娜塔的妈妈，你可能会想：她简直就是个巫婆，存心想淹死娜塔。

But it is always the best way not to be quick to believe bad things about anybody, because most of the time they are not true. That is how it was with Nata's mother. She never tried to catch a girl or boy in her life and she wasn't anybody's darning-needle, and she

couldn't sew up anything even if you held her right up to it.

可是,聪明人是不会轻信谣言的,因为,人们常常无中生有。娜塔的妈妈就是被冤枉的。她一辈子从来没有抓过谁，不管是男孩还是女孩；她也不是谁的缝衣针，就算我们拿着她去缝东西,她也根本缝不了。

And Nata did not drown. She hatched out as well in the water as Lampy did in the ground, or as Luna did on a green oak leaf. It was just the right place for her, and her mother had done her no harm in putting her there.

再说了,娜塔也根本没有被淹死。她好好地在水里孵化呢,就像萤火虫小亮亮在地里孵化、芦娜在绿油油的橡树叶上孵化一样。那里就是最适合她的地方,她的妈妈把她放在那儿,对她没有一点儿害处。

In fact, Nata's mother is known to have done much good in the world—She caught so many flies and mosquitoes that she was sometimes spoken of as a mosquito hawk. She was a great hunter all her days,

and her real name, after she got her wings, was Dragonfly.

其实,大家都知道,娜塔在这个世界上做了很多好事。她抓了那么多苍蝇和蚊子,有时候人们干脆称她为“吃蚊子的鹰”。她毕生都是一个捕蚊高手,有了翅膀之后,她的真名是“蜻蜓”。

When Nata hatched she had no wings and she was called a nymph, which was just a name she had while she lived in the water—the same as Poly was called a caterpillar while she was living on a parsnip plant.

娜塔刚孵化出来时没有翅膀,被叫作若虫,那只是她生活在水里时的名字,就像蝴蝶波丽生活在防风草上时被叫作毛毛虫一样。

Nata, the nymph, looked as much like a grown-up dragonfly as Poly, the caterpillar, looked like a butterfly. Maybe she looked a little bit more like one, for she had six legs from the first to the last, while Poly started out with sixteen and ended up with six. But it would take more than the same number of legs to make Nata, a nymph, look like her mother, a dragonfly. What she did look like, it would be hard to say. Perhaps, more than anything else, she just looked comical.

毛毛虫波丽的模样有多像蝴蝶，若虫娜塔的模样就有多像成年蜻蜓。也许，若虫娜塔更像蜻蜓一点，因为她自始至终就六条腿，而毛毛虫波丽刚开始是十六条腿，后来就只剩六条腿了。可是，若虫娜塔长得像她的蜻蜓妈妈，不仅仅是因为她们的腿一样多。那么，她到底哪里长得像妈妈呢？这个很难说。也许，最像的就是她那副滑稽样吧？

At first, of course, she was tiny, as all things are that hatch out of small-sized eggs. From time to time, as she grew, she had several different bathing-suits, each one larger than the one before. The first one didn't have a sign of wings on it; but as she grew bigger and changed one bathing suit after another, the signs of wings began to come, and then to grow bigger with each new suit. Not real wings, you know—just signs of them that are called wing-pads, and that look like four little flat flaps on her back.

当然，刚开始的时候，她很小，和所有从小个儿的卵中孵化出来的东西一样。后来，她渐渐长大，这期间，她时不时地换泳衣，换了好几套，一套比一套大。第一套没有翅膀的痕迹，可是，当她越长越大，泳衣换了一套又一套的时候，就出现了翅膀的痕迹，而且，泳衣越大，痕迹也越明显。你知道的，不是真正的翅膀，只不过是一点儿痕迹，被叫作翅芽，看上去像是垂在她背上的四个小薄片。

When she got large enough for a change, a new suit grew under the one she had been wearing, and then the old one ripped far enough for her to step out of the tear. She did this under water, so you may think she must have been washed very clean all the time. But, you see, she waded around where it was muddy so much that it was hard to tell what was mud and what was Nata, the mud-covered nymph; just as it was hard to tell what was sand and what was Carol, the sand-colored locust. So some of the mud would have to be washed off before we could see that Nata was comical.

每次发育到需要换衣服的时候，在她正穿着的泳衣下面就会长出一套新的来，然后，那套旧泳衣就裂开一个大口子，一直裂到脚底，她一迈脚就能从裂开的口子里出来。所有这一切都是在水下完成的，所以，你可能会想,她身上肯定总是洗得干干净净的。可是,你错了。她经常在泥糊糊的地方滚来滚去,弄得浑身是泥,结果,很难分清哪里是泥,哪里是若虫娜塔。就好像很难分清哪里是沙子,哪里是灰头土脸的蚱蜢欢欢一样。所以,得把娜塔身上的泥洗掉一些,我们才能看清她有多滑稽。

To begin with, she wore a mask which covered nearly all of her face, as far up as her feelers in front and as far as her eyes at the side. This mask was as big for her as one would be for you if you make one by holding your arms together in front of you, bent up from the elbow so that you can cover your mouth and nose with your hands. If you do that, you can throw your mask out in front of you so that your face shows, and then draw back again and cover it up. Nata's mask was hinged like that, so that she could throw it out quickly and pull it back. Most of the time it was folded up tightly over her big mouth. It kept the mud out when she dived down head-first.

先让我们来看看她的那副面具。它往上最远到触角,往两边最远到眼睛,差不多把她的整个脸都遮住了。这副面具的大小,你自己比画一下就知道了。伸出双臂,弯肘,用双手蒙住嘴巴和鼻子。哎,对了,就那样。你可以往前拿开你的双手,露出你的脸;你也可以往后收回双手,又蒙住你的脸。娜塔的面具就是这种折叠式的,可以迅速地移开,也可以迅速地蒙上。大多数时候,她的面具都紧紧地折起盖住她那大大的嘴巴。这样,当她头冲下潜入水中的时候,嘴巴里就不会进泥了。

It did something else, too. For Nata was a hunter, and her mask was a little trap she carried around with her all the time. She hunted in the water as her mother hunted in the air, and as Lampy hunted in the ground; only when her mother and Lampy hunted, they went after things. Nata didn't have to. She sat still with her trap folded over her face, and when a wriggler (that is the name of a young mosquito before it has wings) or something like that came near, she would just throw out her trap and pull her food right up to her mouth. It was a pretty good trap; it never got out of order, and every new bathing suit had hitched to it a new mask that was of the same sort only bigger than the one before.

她的这个面具还另有用途呢。娜塔是个捕捉能手,面具就是她的小捕捉器,她总是把面具走到哪里就带到

Nata was a hunter.
娜塔是个捕捉能手。

哪里。她在水中捕食，就像她的妈妈在空中捕食，萤火虫小亮亮在地上捕食一样。只不过她妈妈和萤火虫小亮亮捕食的时候需要追着目标跑，而娜塔不用。她把捕捉器叠起来放在脸上，静静地坐在那里。如果有孑孓（还没有长翅膀的小蚊子）或者类似的小东西靠近时，她只需伸出她的捕捉器，把它直接拽到嘴里。小捕捉器很好使，从不出毛病。每套新泳衣上都有一个新面具，和前一个一样，只不过要大一些。

Her feelers were just above the edge of the mask, so that she could move them around in the water. They were slender, and the seven joints in them were all very short so that she could not feel very far.

她的触角就在面具的边上，可以在水里四处打探。由于触角纤细，七个关节也都很短，所以，她打探不了多远。

When the top of her head showed, the two eyes stuck out round like those of a frog. This made her look very wide-awake indeed. But when the front and side of her head showed, her eyes were different from

any other kind I have ever seen. They had wide stripes of pale green and narrow ones of dark purple, and the stripes were slanted. Now eyes with green and purple stripes are pretty—very pretty; but they are funny, too, and her queer eyes and her queer mask, both together, were what made Nata's head so comical.

等到她露出脑袋顶来的时候,就能看到两只眼睛像青蛙的眼睛那样圆鼓鼓地瞪着，一副非常警觉的样子。可是,等她露出脑袋前面和侧面的时候,我就发现她的眼睛和我见过的所有其他眼睛都不一样。上面有条纹,宽的是浅绿色的,窄的是紫色的,而且都是斜的。带绿色条纹和紫色条纹的眼睛一般都很漂亮，非常漂亮,不过,也很搞笑。怪怪的眼睛和怪怪的面具,正是这两样加起来,让娜塔看起来这么的滑稽。

If her head was a strange one, so was her tail, for that was where the breathing tube was.

如果说她的脑袋很奇怪，那么她的尾巴也很奇怪,因为她的呼吸管在那里。

Of course no little Hexapod ever breathes through

his head. Not Van or Gryl or Old Bumble or any of them. Grown-up ones, by the time they have wings, breathe through a row of little holes on each side of their bodies. Most baby Hexapods breathe the same way, but Nata, the nymph, was different. She had a breathing tube with an opening at the tip of her tail. So it was almost as if she had a mouth at each end of her body—one to eat with and one to breathe with.

当然，没有哪个六脚节足小昆虫是用脑袋来呼吸的。蝴蝶帆儿不是，蟋蟀唧唧不是，黄蜂老奔奔不是，他们都不是。成年昆虫长出翅膀后，就用身体两侧的一排小孔来呼吸。大部分六脚节足昆虫宝宝的呼吸方式都一样，但是若虫娜塔除外。她有一根开口在尾巴尖的呼吸管，所以，几乎像是她有两个嘴巴，身体两头一头一个。一个用来吃饭，而另一个则用来呼吸。

You know that most animals that live in the water have gills to breathe with. Nata did, too, and her gills were on the inside of her breathing tube.

你知道大部分生活在水里的动物都是用鳃呼吸的。娜塔也有鳃。她的鳃就长在她的呼吸管内侧。

When you breathe, you have to get oxygen out of the air. When Nata breathed, she had to get some oxygen that was in the water. So she would draw the water into her tube as we draw air into our lungs, and then push it out again as we push out the air.

你呼吸的时候，得从空气中吸取氧气。娜塔呼吸的时候，也得吸取一些水里的氧气。所以她会把水吸到呼吸管里，就跟我们把空气吸到肺里一样，然后把水排出去，就跟我们把空气排出去一样。

So now you can see why, when Nata used to dive down head-first, she left the tip of her tail up out of the mud where there was some water. If she had stayed tail down and head up, she would have smothered.

所以，现在你明白了吧？为什么当初娜塔头冲下潜入水中的时候，要把尾巴尖露在泥水的外面。如果尾巴朝下、头朝上，她会窒息的。

Nata liked diving and did a great deal of it. Sometimes she would stand on her head down deep in the mud for a long time. It seemed to be one of her ways of hiding.

娜塔喜欢潜水,而且经常潜水。有时候,她会倒立在深深的泥水里,一待就是很长时间。这似乎是她的一种隐身术。

When she hid another way, she would lie down in the mud and then kick with all six of her little feet until she scratched out a hole in the mud, and her body would sink down deeper and deeper until she would be all covered up except her eyes and the top of her trap at her head-end and her breathing hole at her tail-end. While she was hiding that way, she was usually hunting.

她的另外一种隐身术是躺在泥里，六条腿齐上阵，使劲地踢,直到在泥里踢出一个洞为止。她的身体也会慢慢沉到洞里，越沉越深，直到整个儿埋在泥下面,只露出眼睛和头下面的那点小捕捉器,还有尾巴尖上的呼吸孔。她这样隐藏时,多半是在捕捉什么东西。

The mud Nata played in and hunted in was at the bottom of a pool a little off at one side in Shanty Creek. Now, you can tell from this something about what part of the country she lived in, because if it had

Nat was Nata's mate.
奈特是娜塔的伴儿。

been in some other places the name would have been Shanty Stream or Shanty Brook.

娜塔是在一个水塘底的淤泥里玩耍和捕食的。水塘的一边不远处是一个叫“木屋溪”的地方。现在,从这一点你就能判断出她住在乡下的什么地方了吧? 因为如果是别的什么地方,那个地方就要叫“木屋河”或者“木屋江”了。

Nata did not stay in the mud at the bottom of the pool all the time. Now and then, though not often, she would walk out near the edge where the water was not deep, and put the mouth of her breathing tube up into the air; and it would look as if she could breathe that way, too. And sometimes she would run about among the wet dead grass-stems and other rubbish in the water near the edge.

娜塔并没有一直待在水塘底的淤泥里。偶尔,她会走到塘边水浅的地方,把她的呼吸管口伸出水面,好像那样她也能呼吸似的。有时候,她会在池塘边那些湿漉漉的枯草丛中和其他蚊蝇孳生的地方跑来跑去。

One day she did this after a heavy rain, when there was a flood that had filled a hollow next the pool she lived in. The water spilled over from the pool into the hollow, and Nata happened to walk into the hollow where there were some good things to hunt that had gone over with the water.

一天，一场大雨之后，她就是这么过的。大水把水塘旁边的一个水坑给填满了。池塘里的水溢出来，流到了那个坑里，有一些好东西也随着流了进来。那天，她碰巧也走进了坑里，刚好就赶上了。

That did very well for a day or two, and Nata stayed. She stayed so long that the water went lower and lower, until it no longer spilled over from the pool into the hollow.

头一两天，她收获颇丰，所以，就一直待在那个坑里没出来。可是后来，水塘的水位越来越低，最后，水就不再从水塘溢到坑里了。

Now, the hollow was small, and it did not take her many days to trap all the good things there were in it. After that, of course, there was nothing for her to do

but wait; for she did not know anything about going from the hollow up the dry bank and over into the pool again. She couldn't think things out like that. She had a brain in her little head, too, and a wonderful one it was; but it didn't help her back into the pool.

水坑小了。几天工夫她就捉完了里面所有的好东西。当然,从此,她除了等待没什么事可干了。因为,她不知道她还能从坑里上来,越过干坑堤,重新返回水塘。她产生不了这样的念头。她的小脑袋瓜里也有大脑,很棒的大脑,可是,这个大脑却无助于她返回水塘。

It could help her wait, though, and that is what she did. She waited for seven days, and nothing—not so much as one mosquito wriggler—got into her trap. And all this time the water was sinking lower and lower in the little hollow, until by the end of the week there was not more than a cupful left. Then it rained until there was as much as would fill a dishpan.

不过,这却有助于她耐心等待。她就一直在等待,等了七天,连一只蚊子孑孓都没有捉到。这期间,小坑里的水位越来越低,一周之后,坑里存的水连一杯都不到了。就在这个时候,开始下雨了,一直下到坑里的水差不多能装满一个洗碗盆了。

Nata still waited, with her trap held up closed about her hungry mouth, for days and days and nights and nights, until another week went by, and she had nothing to eat even then.

By that time the water was gone from the hollow. There wasn't enough left to fill a thimble!

The dirt at the bottom was still soft and wet, though, and there Nata stayed, all alone and very hungry.

娜塔还在等，合起小捕捉器举在饥饿的嘴巴跟前。一天又一天，一夜又一夜。又一个星期过去了，她依然没有东西可吃。

那个时候，坑里的水所剩无几，连一个顶针都装不满了。

不过，坑底的土依然松软、潮湿。娜塔就待在那里，孤立无援、饥肠辘辘。

Poor little Nata, waiting in the mud! Four days went by while she had just air to draw into her breathing tube, for all the water was gone. So there

were four more days to add to the other fourteen without food!

可怜的小娜塔，在泥里苦苦等待着！又过去了四天，这期间，她只能用呼吸管吸一点儿空气，因为一点儿水都没有了。也就是说忍饥挨饿十四天之后，她又有四天没有吃东西！

And Nata still waited. Do you wonder how she lived so long? Perhaps it was partly because she didn't tire herself all out fretting. She waited quietly most of the time, and rested. Partly too, it was because she was strong and because she was a hunter, and very often hunters have to wait a long time between meals. But eighteen days was longer than usual, and unless something happened to get Nata back into her pool, she would—Oh dear, oh dear! What would become of her?

娜塔还在等。你想知道她怎么坚持了这么长时间吗？也许是因为她没有烦躁不安消耗体力。大部分时间，她都在静静地待着休息。也许是因为她比较强壮，因为她要捕猎。猎人经常是吃了上顿后要等很长时间才能有下顿的。可是，十八天的时间有点太长了。如果不发生点什么事把她带回到水塘里，她就…… 噢，天哪，噢，天哪！那她会怎么样呢？

But something did happen. It was rain! There was thunder and lightning with it, but Nata didn't mind that. And there was a flood—a nice big one that spilled the pool over into the hollow.

不过，真的发生了点事情。下雨了！电闪雷鸣、大雨滂沱，然而，娜塔可不介意。发大水了，很及时的大水，从池塘里溢出来流进了坑里。

In a little while after that, Nata found her way back to her old hunting-ground in the mud at the bottom of the pool. Perhaps she felt the motion of things swimming about in the pool. Perhaps she smelled something that tempted her back. Perhaps she just went walking around and happened to get home again as she happened to get lost.

没过多久，娜塔终于回到了她那位于塘底的淤泥捕捉场。也许是她感觉到池塘里有东西在游动。也许是她闻到的什么气味把她吸引回来了。也许是她四处闲逛的时候，碰巧就回了家，就跟当初她碰巧迷了路一样。

Whatever way it was, she had a good time all the rest of the days she was a nymph. She caught plenty of food in her trap—nice tender juicy food. She played in the mud, sometimes diving down in it head first, and sometimes kicking it out from under her so that the trap on her head was where she could use it as well as the opening of the breathing tube in her tail.

不管是怎么回事，总之，若虫时期，她后来的日子过得都很开心。她的小捕捉器收获甚丰，都是可口、鲜嫩、汁液丰富的美食。她在淤泥里尽情玩耍，时而头冲下潜到泥里，时而又把泥从身子底下踢开，让脑袋上的小捕捉器和尾巴尖上的呼吸管的管口各就各位，随时准备工作。

Then one day, right in the midst of all this fun of diving and wading and hunting—one day she stopped doing the things that she had done all her life, and walked up out of the water to the shore of the pool, and climbed a little way up the stem of a plant, and then pretty soon all there was left of Nata, the nymph, was an empty bathing suit, just her shape, clinging to the plant.

后来有一天，她在淤泥里进进出出、滚来滚去、追逐美食，玩得正开心的时候，突然间，就停了下来，终止了这些她一直都在做的事。她从水里走出来，来到水塘堤岸上的一棵植物跟前，顺着茎秆往上爬了一段，没过多久，那茎秆上就只剩下若虫娜塔的泳衣了，空空的一个外壳，还紧紧地抓着茎秆。

Nata the Nymph is no more. She has gone, leaving her water-clothes all whole and good except the opening at the back where she pulled herself out of them. They look as if they were left waiting for her to come and slip into them again.

若虫娜塔消失了。她走了，留下了她的泳衣。除了后背上那个她用来脱身的开口之外，泳衣依然完好，看样子，好像是在等她回来再钻进去一样。

When will she come back and live in her pool?

什么时候她才能回来，继续水塘里的生活呢？

Ah, Nata can never do that. Look, and you will know why!

Well—She came out of Nata's bathing suit.
对了,从娜塔的泳衣里出来的就是她。

哎呀！告诉你吧，娜塔永远不会回来了。往那边看，你就知道这是为什么了！

See—flying along the bank of Shanty Creek, a beautiful creature with four wonderful strong wings and with eyes that gleam like living jewels! Oh, now she has stopped and is resting in the sunshine near Nat.

瞧！沿着“木屋溪”的溪岸，一个美丽的小家伙正翩翩起舞。她的四只翅膀精美绝伦、强劲有力；眼睛闪闪发光，像熠熠生辉的有生命的宝石！哇，瞧，现在，她收起了翅膀，正沐浴着阳光在“娜塔”旁边休息呢。

Well—She came out of Nata's bathing suit. Do you think she will ever need it again?

对了，从娜塔的泳衣里出来的就是她。你想她还会需要那身泳衣吗？

7

Lampy's Fourth O'July
萤火虫小亮亮的国庆节

No one had ever given Lampy so much as a penny. He had never earned any money. He had never found any. Here it was the Fourth o' July, and he couldn't buy even one fire-cracker! Not any of his little gang of Will-o'-the-Wisps had fire-crackers either, so you might think that they were going to have a dull time of it that night.

谁也没给过小亮亮钱,一分钱都没给过。他也从来没有挣过钱，更没有捡到过钱。现在是七月四号国庆节,他连一个鞭炮都买不起！他的"小鬼火帮"里,也没有哪个能买得起鞭炮。所以,你可能以为那个晚上他们会过得很无聊吧？

For the Will-o'-the-Wisps had a habit, during the summer, of meeting every evening about the time the stars came out, and having a frolic together. They never did any harm in the world. If the policeman saw them darting about after the curfew rang, he just smiled and never thought of such a thing as sending them to bed.

夏天的时候，这些鬼火帮的小家伙们有个习惯，每天晚上星星出来的时候，他们都要聚在一起嬉戏玩耍。他们从不做坏事。宵禁铃拉响之后，要是警察看到他们还在东游西窜，也只会笑笑而已，绝不会想到要赶他们回去睡觉。

Why, indeed, should Lampy go to bed by night? He rested all day, like a little owl. Of course it is all right for some folk to sleep while it is dark. Van and Poly and Jack and Nata and Carol are awake all day, and so they need to rest at night. They are made that way, just as chickadees and robins are. Some flowers are like that, too, and shut up their pretty eyes as soon

as it grows dark.

的确,晚上小亮亮干吗要睡觉呢?他像小猫头鹰一样,白天可是一直在休息呀。当然了,有些人天一黑就睡觉。蝴蝶帆儿、蝴蝶波丽、角蝉杰克和蜻蜓娜塔,还有蚱蜢欢欢,他们白天兴奋了一天,所以,晚上需要休息。他们生来如此,就像山雀和知更鸟一样。有些花儿也是这样的,天一黑,就闭上美丽的眼睛。

But Lampy! My, no! Miss the night-time? Sleep while the Big Dipper in the sky whirled slowly around the steady North Star? Doze while the evening primrose opened its lovely yellow cups of nectar and filled the air with sweet smells? Nap while the new moon sailed the skies like a little boat?

可他是小亮亮!要他错过夜晚的大好时光?当天上的北斗七星围着纹丝不动的北极星旋转时，让他睡觉?当夜来香打开黄色的蜜杯，夜空中弥漫着浓郁的芬芳时,让他打盹?当新月像一只小船,在夜空中徐徐航行时,要他小睡?噢,天哪,这可不行!

Why, even the bat, the cunning flying mouse,

came out to enjoy the night air. Gryl was fiddling merrily in the fields. Luna was floating, like a fairy robed in white and green, through the woodland.

再说了,连蝙蝠,狡猾的小飞鼠也都出来尽情享受夜晚的空气了;田野里,蟋蟀唧唧欢快地拉起了小提琴;月形天蚕蛾卢娜也在树林里飘然而行,像穿着白绿两色长袍的仙女。

Oh, no! Lampy could not miss the night. Not that he listened to Gryl's music, or smelled the evening primrose, or watched the beautiful Luna, or even looked at the sky! But the night was good to him even without these joys. He was a part of it, and, besides, the whole gang of Will-o'-the-Wisps would be out for a Fourth-o'-July dance.

噢,不!小亮亮可不能错过美妙的夜晚!不是因为她听到了蟋蟀唧唧的音乐,也不是因为他闻到了夜来香的芬芳,也不是因为他看到了漂亮的月形天蚕蛾卢娜,更不是因为他仰望夜空看到了群星和月亮。即使没有这些令人赏心悦目的事情,他也喜欢夜晚。因为,他是夜的一部分,再说了,他们"鬼火帮"还要全体出动去参加国庆舞会呢。

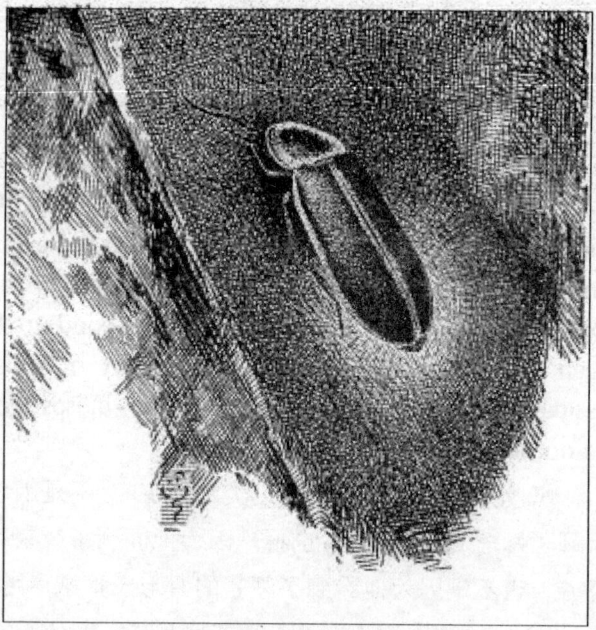

Oho! Here was Lampy.
哦哈！小亮亮来了。

And the dance, just think of it, the dance was going to be in the air!

A Fourth-o'-July dance without fireworks, though—dearie, dearie, me!

这个舞会，哇，想想吧，这个舞会将在空中举行！

一场没有烟花的国庆舞会，噢，天哪，天哪，我的天哪！

Oho! Here was Lampy! He had crept out of bed at dusk, had lifted his upper wings, which covered the ones he used for flying, had shaken the wrinkles out of his thin under ones, and here he was, ready for the dance. And here was Jack-o'-Lantern, and here was Wah-wah-taysee, and here were Tinker Bell and Star-Light and Eye-Bright and Ray and Beam and Flash and Gleam-here were, indeed, the whole gang of Will-o'-the-Wisps! And every one of them carried a candle.

哦哈！小亮亮来了！黄昏时分，他从床上爬起来，把盖在飞翅上面的上翅抬起来，抖平薄薄的底翅。现在，他来了，一切就绪，只待开跳。南南灯来了，哇哇熄来了，叮叮当、星星光、亮亮眼、光线线、光柱柱、小闪闪、小烁烁都来了。真的，整个“鬼火帮”都来了！大家都带着一只烛灯。

Over the marsh the mist is white,
The owl is calling through the night;
While, like a flock of dancing stars,
The Will-o'-the-Wisps are taking flight.

沼泽上方白雾蒙蒙，
猫头鹰的叫声划破夜空；
吓得“小鬼火”四处奔逃，
仿佛一群蹦蹦跳跳的小星星。

It was a dance to be happy about, for there was a time when none of these Will-o'-the-Wisps had any wings; and, as their six little tiptoes were not made for dancing, they had kept very close to the ground. So close, indeed, that they had lived right in it all the while they were growing up.

这个舞会真令人兴奋！曾经有一段时间，“小鬼火帮”谁也没有翅膀。另外，由于六个小脚尖不适合跳舞，他们只能紧紧地扒住地面。他们扒得太紧了，长大之前，他们简直就是整个儿直接生活在地上。

No, Lampy had not always been flying through the air at night and resting wherever he happened to be by day. He had stayed where his mother had put him.

那时，小亮亮可不是每天晚上都在夜空中飞来飞去，也不是大白天想在哪儿休息就在哪儿休息。他很听话，妈妈把他放在哪儿，他就待在哪儿。

His mother, whose name was Firefly, wore over her head something that looked like a flat red-and-black hat, and her eyes were hidden under the broad yellow brim. She had a candle very much like Lampy's, only not quite the same in shape; and her wings were dark, with a light yellow edge all round. Like Lampy, she had rested by day, and had flown with her kind at night.

他的妈妈叫“萤火虫”，头上好像戴着一顶红黑两色的平顶帽，黄色帽檐宽宽的，把眼睛都给遮住了。她也有一根蜡烛，和小亮亮的差不多，不过形状不太一样。她的翅膀颜色很深，镶着一圈淡黄的边。和小亮亮一样，她也是白天休息，晚上跟她的同伴一起到处飞。

Her candle, it is thought, was a signal to other firefly beetles to come and join the dance, for they flew in large flocks near the swamps and over the low meadows. Her bright light did not give a steady glow like the North Star, but flashed and then went out and flashed again; as you can make a little pocket-light do by pressing it. So she could give "wig-wag" signals to her friends; and, as it was night, of course a flashing candle was better than a flag.

据说，她的小蜡烛是信号灯，可以召唤其他萤火虫来参加舞会，因为他们喜欢聚在沼泽地和矮草地的上空翩翩起舞。她那盏小灯发出的光不如北极星的光稳定，但是，却一闪一闪的，就像小手电筒那样，你一摁它就亮，再一摁它就灭。所以，她能给朋友们发信号，当然，因为是晚上，所以闪动的烛光要比信号旗管用。

It is thought, too, that her candle was a warning to the night birds. It showed them just where she was, and they needn't swallow her up if they didn't like the taste of fireflies; just as Poly's bright dress showed the

day folk where she was, so that they need not meet her unless they wished to.

据说，她的小蜡烛还是警示灯，可以让夜鸟清楚她的位置，让她们知道如果不喜欢萤火虫的味道，就不要一口吞了她。这就像蝴蝶波丽，大白天穿着鲜艳的衣服，就是为了让大家知道自己的位置，这样，如果不想和她打照面，完全可以躲得远远儿的。

Well, Firefly did not spend all the time in dancing at night and in resting when it was day. She had some eggs to take care of, and no matter how they may have looked to others, they were as good as gold to her.

She buried her treasure—Lampy and the other little eggs. So their cradles were right in the dirt.

不过，萤火虫并不是整个晚上都跳舞，也不是一到白天就睡觉。她还有一些卵需要照顾。不管在别人眼里他们是怎么样的，对她来说，他们就像金子一样宝贵。

她把她的宝贝——小亮亮和别的几个小卵埋了起来。所以，他们的摇篮就在土里。

That was a handy place to be in, as Lampy found

when he hatched and crept out of the egg, for he was born a hunter, and had to catch every bit of his own food.

小亮亮孵化好，从卵里爬出来之后，才发现待在这里可真方便，因为他生来就是个辛苦命，所有的食物都得自己去捕去捉。

Underground there were as many little things that Lampy liked to eat as there are on top of the ground for Biddy and her chicks, when they go hunting, or in the air for Nata, when she is grown up and flies about, or in the water for her, when she is a nymph and hides at the bottom of the pond.

地底下蕴藏着丰富的小东西，都是小亮亮爱吃的。到底有多少呢？这么说吧，地上有多少母鸡叨叨和她的宝宝们喜欢吃的东西，地下就有多少小亮亮爱吃的东西；天上有多少蜻蜓娜塔喜欢吃的东西，地下就有多少小亮亮爱吃的东西；水里有多少若虫娜塔喜欢吃的东西，地下就有多少小亮亮爱吃的东西。

Food enough in the world for everybody, if

Like a flock of dancing stars the Will-o' -the-Wisps are taking flight.
沼泽上方白雾蒙蒙,猫头鹰的叫声划破夜空。

everybody goes about getting it in the right way; and the right way for Lampy was hunting in the ground!

这个世界上的食物足够所有的人吃,只要大家取食得当。对小亮亮来说,在地下捉东西就是他得当的取食之道。

So he dug little holes wherever he went, and made little caves and crept through them, and hunted and hunted and hunted all the time he was growing up. There, too, underground in his own dark hole, he changed his clothes when they felt too tight.

所以,没长大之前,他每到一处,都会挖小洞,做成小巢穴,然后爬过去,捉呀,捉呀,一直捉个不停。也就是在那里,在他自己那黑乎乎的地下小洞穴里,他换掉了一件件紧得无法再穿的衣服。

He was a queer-looking little thing, with a small head, and then a lot of joints for the rest of his body that were very much alike, except that the tail one had a little tool on it, and the three nearest his head had each a pair of short legs.

他是个长相奇怪的小家伙。小小的脑袋,身体有好多节，差不多都一样，只不过尾巴上的那个节带一个小工具，接近脑袋的那三个节各有一双短短的腿。

But one day he stopped looking like that. This was after he slipped out of his last suit of jointed clothes and went to sleep as soon as they were off. For little Lampy had grown to the time when he was to hunt no more through holes in the ground. He was going to sleep, and then he was going to waken and have wings to fly with and a candle to flash in the night-air. Think of that!

可是，突然有一天，他变了模样。这是他脱掉最后一套关节服，倒头入睡之后发生的事情。因为，小亮亮已经长大，不需要再在地下的那些洞里捕食了。他要睡觉，然后，他要醒来，长出几只能飞的翅膀和一个能在夜空中闪烁的小烛灯。想想吧，太令人神往了！

So he slept in the earth, and his bed under him was good clean dirt, and his blanket over him was good clean dirt, too.

Then, when it was the right time in the summer, he climbed up out of the ground. And that evening he

flew over the low meadow to the marsh. There he found Jack-o'-Lantern and Wah-wah-taysee and Tinker Bell and Star-Light and Eye-Bright and Ray and Beam and Flash and Gleam.

所以,他睡在地下。身下的床是上好的净土,身上的毯子也是上好的净土。

接下来,在夏天时机成熟的时候,他爬出了地面。那天晚上,他飞过矮草地,来到了沼泽地。在那里,他碰到了哇哇熄、叮叮当、星星光、亮亮眼、光线线、光柱柱、小闪闪和小烁烁。

They all danced in the air with their wings; and their candles, twinkling in the dark, were more beautiful than any other lights that could be seen—except the stars.

他们都扇动着翅膀在空中飞舞。身上的小烛灯在黑暗中一闪一闪的,比所有能看见的灯都美丽,当然,星星除外。

Oh, is it, then, a fairy sprite,
That frolics with the elves at night?

The Will-o'-the-Wisps, on dancing wings,
But wake the dark with their delight.

哦,难道是仙界的小精灵,
在夜色里与捣蛋鬼们嬉闹?
荧光闪闪乘着舞动的翅膀,
黑夜也因他们的欢乐而睁开了眼睛。

Way, way off over the city, the sky-rockets and other fireworks showed where a crowd of men were having their Fourth o' July, too; but, do you know, with all their money, they had not been able to buy anything so lovely as the dance of the Will-o'-the-Wisps over the marsh!

很远、很远的城市上空,流星焰火和其他焰火五彩缤纷,那里的人们也在欢度国庆节。可是,你知道吗?就算他们花光所有的钱,也买不到沼泽地空中舞会上"小鬼火帮"们的快乐!

For Lampy's candle was a secret, and no man has ever been able to make one like it. Not in this country or way over on the other side of the world in China,

or anywhere, has any man ever made a candle such as Lampy and the other Willo'-the-Wisps carried that night as they danced in the air above the marsh.

因为,小亮亮的烛灯是个秘密,还没有人能制造出这样的一只烛灯。就是小亮亮和其他“鬼火帮”小家伙们在沼泽上空跳舞时拿着的那种烛灯。我们这儿没有,地球那一端的中国也没有,别的地方也没有。

The most wonderful thing about it was that it was a cool light. You know that you cannot put your hand into the flame in the fireplace without getting burned. The rays of the sun are warm even when they have come many miles to us. You know that the light from kerosene or alcohol is hot enough to cook your dinner with, if there is an oven over it.

最神奇的是,那烛灯是凉的。你知道的,如果你把手伸进炉火里,肯定会烧伤你的手;太阳光穿越了很远的距离才来到我们这里,但是依然暖洋洋的;你也知道,如果在煤油灯或者酒精灯上架一个锅,其热量足可以用来做饭。

And every one of them carried a candle.
大家每人都带着一只烛灯。

But Lampy's candle was pure light, with no heat. It was ten times as good a light as electricity makes, and fifty times as good as gas; and it was so cool that it did not burn his wings when they were folded right over it. It did not even make the end of his body, where he carried it, one bit warmer than his head.

可是,小亮亮的蜡烛只发光,不发热。它要比电灯好十倍,比油灯好五十倍。他的烛灯凉凉的,翅膀直接叠放在上面,也不会烧坏。烛灯就装在他的尾部,但是,那里一点儿也不比他的头热。

Think of a light like that for a hot summer night! Think of Lampy, without a penny in the bank, owning a candle such as the richest man in the world cannot buy! Think of having a secret that the wisest man never has guessed!

想想吧,夏日炎炎的夜晚,能点这样的一盏灯多好!想想吧,连一分存款都没有的小亮亮,竟然有一盏这样的灯,连世界上最有钱的人都买不到的灯!想想吧,他掌握的秘密,连最聪明的人都没猜出来过!

Mind, I do not say that no one ever will find out the secret of Lampy's candle. Maybe you will yourself, when you are grown up. Who knows? Just because no man or woman ever has, there is no reason to say that no one ever can.

注意！我可没说将来也没人能发现小亮亮蜡烛的秘密。也许等你长大了，你就能。谁知道呢？现在还没有，并不等于将来也不能呀。

If you ever do learn how to make a pure light that is not hot or even warm, what will you do with it ? Will you keep your secret all your days? Will you sell it and be very rich? Will you give it to poor people in hot cities, that they may have a light that is cool? Will you give it to the sick, who should not have a warm light near their heads while they read? Or will you take it as Lampy did, and dance in the night with your friends, making the dark earth beautiful with your candles as the dark sky is beautiful with stars? And will you sing as you go:—

假如你真的掌握了这个秘密，学会如何制作一盏只发光不发热，甚至连一点温度都没有的灯，那么，你会怎么样呢？你会一辈子都守着这个秘密吗？你会卖掉它,然后变成富人吗?你会把它送给炎热城市里的那些穷人,让他们拥有一盏凉爽的灯吗?你会把它送给那些头顶附近不能有热光的病人,让他们也能享受阅读的快乐吗?你会像小亮亮那样,拿着它和朋友们一起在夜色里翩翩起舞,让漆黑的大地因你们的烛灯而熠熠生辉,就像茫茫的夜空因群星而璀璨夺目吗? 当你手舞足蹈的时候,你会唱起这首歌吗?

"The flowers of dusk are gleaming bright,
And give their sweetness to the night;
While day-folk sleep the dark away,
We dance by pretty candle-light! "

“黄昏之花闪闪发光,
夜色里弥漫着芬芳;
日出而作的人们在漫漫长夜里睡去,
我们在美丽的烛光中翩翩起舞!”

8

Carol

蚱蜢欢欢

Her real name was Carolina Grasshopper, but we will call her Carol for short.

Carol was sitting in the sandy path, so near the road that, when any one passed, she had to move out of the way. This did not matter at all, for one place in the sand was as good as another to Carol.

她的真名是蚱蜢卡罗琳娜，但是，为了简便，我们叫她欢欢。

欢欢正坐在一条沙土小路上。她离路边太近了，所以，只要有人经过，她就得让路。不过，欢欢一点儿也不介意，对她来说，只要是在沙地上，哪里都一样。

It was noon -time of the hottest day of the summer, and the sun shone down on her back and head, and the sand she was sitting on was almost hot enough to hurt your hand. It was the sort of day, in fact, when a person, if he does not put a damp cloth under his hat, is likely to have a sunstroke.

But Carol did not even have a hat on.

那天中午是夏天里最热的一天,太阳照着她的脊背和脑袋。身下的沙子热得几乎烫手。这样的大热天,如果不在帽子里垫一块湿布,很容易中暑。

可是,欢欢连帽子都没戴。

As for leaving that hot sand and having a nice cool wade in the little brook that sang "Bip -po bap -po" down through the shady woods to the river not far away—why Carol would never have done such a thing in the world! Had her mother not sat in the sand by the side of the road one year ago; and had not her two grand -mothers done the same thing two years ago? And had her father and her two grandfathers ever found any better place than sand in the sun on a hot summer day? What was good enough for them was good enough for her.

It was noontide of the hottest day of the summer.
那是夏天里最炎热一天的正午时分。

附近有一条潺潺的小溪，流过浓荫茂密的树林，通向不远处的一条河里。欢欢完全可以离开滚烫的沙地，去小溪里蹚水乘凉呀。可是，她死也不会做这样的事情！一年前，她的妈妈就是这样坐在路边沙地上的；两年前，她的外婆和奶奶也是这样；炎炎夏日，她的爸爸、她的外公和爷爷也没找到比烈日暴晒下的沙地更好的地方。所以，他们的好地方，也是她的好地方。

So there she sat, looking about the color of a tiny heap of sand, just as her mother and grandmother and great-grandmother and great-great-grandmother, and all the rest as far back as you can count, had sat and looked before her. You might say it was a habit of her family.

所以，她坐在那里，就像一个小沙丘。她的妈妈、妈妈的妈妈、妈妈的妈妈的妈妈、妈妈的妈妈的妈妈的妈妈，总之，再往前数多少辈，都这样。可以说这是她们的家族习惯。

Now Carol had two little fans, tucked down, one on each side of her. They were pretty when they were

spread out—dull soft black, with yellow borders.

欢欢身上有两把小扇子，收拢起来，放在身体两边，一边一把。小扇子打开的时候，很漂亮。不显山不露水的黑色，柔柔的，还镶着黄边。

I once knew a girl who made one the same size out of black and yellow tissue-paper for her Doll Jane, and it was as cunning a fan as you could wish to see at a doll party.

以前，我认识一个女孩，她用黑黄两色的包装纸给她的洋娃娃做了一个扇子，大小和欢欢的一样，精巧可爱，只有在洋娃娃派对上才能看到。

But Carol was not at a party, and she was not using her fans. She had them all folded neatly under their brown covers, which kept them from getting worn and torn. She never once fanned herself with either of them!

可是，欢欢不是在参加派对，所以，她并未使用她的扇子。她把它们整整齐齐地叠起来，收在褐色的罩子底下，免得把它们弄旧、弄破了。她从来没用这两把扇子

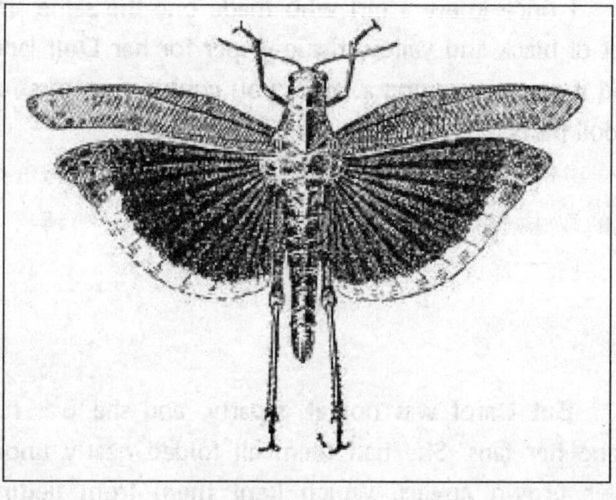

Now Carol had two little fans.
现在欢欢身上有两把小扇子。

扇过风，一次也没有。

With these thin black-and-yellow dainty things tucked out of sight beneath the tough covers that made a sort of roof over her back, there was nothing to show where Carol sat. A little sand-colored grasshopper sitting on the sand—that was all.

罩子很粗糙，像个屋顶一样扣在她背上，黑黄两色薄薄的扇子，虽然精致可爱，但是，被罩子遮起来了，根本看不到她坐在什么上面。只能看见一个土不拉叽的蚱蜢坐在沙土里，仅此而已。

Why, you could have stepped on her easily and never have known it! That is, you could, if she had not seen you coming; for Carol's eyes were very quick to see anything that moved toward her. If things stood still like trees, she did not bother about them, for nothing could catch her unless it came nearer and nearer and nearer. But if old Rover pattered down the road, or a bird flew close overhead and made a shadow, or a child ran by, then quick as a flash Carol would lift her fan-covers out of the way and spread her fans like

little sails and fly off with them. For her black-and-yellow fans were wings!

这么说，你很可能踩到她身上，却永远也不知道！也就是说，如果她没有看到你过来的话，这完全有可能发生。可是，欢欢的眼睛很尖，只要有东西朝她移动，她都能看见。而对于树那样站着不动的东西，她就懒得理。除非是一点儿一点儿慢慢地靠近，否则就甭想逮住她。如果流浪汉老罗佛从那边啪嗒、啪嗒地走过来，或者一只鸟飞到她的上空，弄出一点儿阴影，或者一个孩子从她身边跑过，她都会闪电般地把她的扇子罩掀起来，扬帆一般张开她的小扇子，振翅而去。原来，那黑黄两色的扇子就是她的翅膀！

And when you saw her fly away, perhaps you thought she was a butterfly. Maybe you thought she was a butterfly like Van, with dark wings and yellow borders. Maybe you were the little girl or boy who chased after her along the road, and looked right at her and didn't see her at all; for you were trying to find something with black-and-yellow wings, and Carol was just a sand-colored grasshopper sitting in the sand.

看到她飞走，可能你会以为她是只蝴蝶。也许，她那黑色的翅膀和翅膀上的黄边，会让你以为她就是帆儿那样的蝴蝶。也许，你就是那个小男孩或者小女孩，一路在她身后追逐，目光明明在她的身上，却对她视而不见。因为你找的是长着黑黄两色翅膀的蝴蝶，而欢欢却只不过是坐在沙地上的一只土不拉叽的蚱蜢而已。

Well, if it was not you, it was some other girl and boy; and Carol led them half a mile along the roadside, fooling them time after time by showing them her broad wings in flight and then quickly hiding them under their covers, when she stopped to rest. And by-and-by the children grew so tired that they went off into the woods and found the brook, and waded there until they were cool.

哦，如果不是你，那就是别的小女孩和小男孩。欢欢领着他们跑了半英里，让他们上了好多次当。她一会儿张开她的宽翅膀飞呀飞，一会儿又迅速地把翅膀收起来放在罩子下面，停下来休息。没追多久，孩子们就烦了。他们走进了路边的树林里，发现了小溪，于是就开始蹚水玩，直到浑身凉爽。

On and on flew Carol, a little way at a time, now this way and now that, as dogs and children and horses and men sometimes came up the road and sometimes down. And once Carol was scared into the woods by a funny old cow with a crumpled horn, and the first thing she knew, she was sitting on some pretty leaves on the cool damp ground. But her brown sand-colored body showed very plainly on the green, and she didn't like the shady woods, and when she started for the roadside, she bumped up against some bent grass-stems and fell down. Then she tried jumping out, for—two of her legs were big and strong for hopping; but she kept blundering against stems that were in the way. At last, she crept along more slowly without trying to fly or jump, and so got back to the roadside, where, although many things passed by, none could see her in the sand; and where, although she had often to hop and fly, there were no grass-stems above her to get tangled in.

欢欢飞呀飞，不停地往前飞。每次只飞一点点，一会儿往这边飞，一会儿又往那边飞。因为，总有狗呀、马呀、人呀的，不是从路那边过来，就是从路这边过去。有一次，一头犄角皱巴巴的老母牛走过来，那样子很古怪，把欢欢吓蒙了。等她清醒过来的时候，才发现自己躲在了树林里，周围是冰凉潮湿的地面，而她就坐在一

Perhaps you thought she was a butterfly.
说不定你会把她当成一只蝴蝶。

堆漂亮的树叶上。可是,她沙土色的身体,颜色很深,在郁郁葱葱的绿色堆里很扎眼。再说,她也不喜欢浓荫茂密的树林。于是,她起身往路边飞,可是,刚一飞,就撞在了弯着腰的草茎身上,一下子摔到了地上。于是,她又试着往外跳。因为她的两条后腿粗壮有力,适合跳跃。可是,总是有树干或者草茎之类的东西挡她的路,让她跌跌撞撞,四处碰壁。最后,她打消了飞回去或者跳回去的念头,一路爬回了路边。那里虽然熙熙攘攘,但是,她在沙土里,谁也看不见;虽然在那里她也经常得蹦蹦跳跳、飞来飞去,但是,却不会被讨厌的草茎围困。

But you mustn't laugh too much at little Carol, baking herself in the sunshine. If you had spent the winter where she did, perhaps it would take you all summer to get warm, too.

不过你可千万别起劲儿地笑话欢欢在烈日下暴烤自己,要是你也在她过冬的地方过冬,也许,你也会用整个夏天来暖和身子。

For Carol's mother—just think of it! —poked every

egg she had down as far as she could reach into the ground, and poured over them a sort of froth that hardened about them. Carol herself was the very last egg her mother pushed down, so that she was on top of her brother and sister eggs, which lay under her in slanting rows. But even if she was on top, she had to stay there all winter long, and the ground froze as solid as a cake of ice, and part of the time the weather was colder than zero; and though the snow lay over her like a thick blanket, it was a cold bed for all that, and it is no wonder that Carol didn't hatch for months and months.

仔细想想你就明白了。欢欢的妈妈把她所有的卵都捅到地底下放起来，能捅多深就捅多深。然后，在他们身上倒了一种泡沫，泡沫变硬把他们包了起来。欢欢是妈妈捅进地下的最后一颗卵。所以，她躺在她的兄弟姐妹上面，而她的兄弟姐妹则斜排着躺在她的下面。可是，即使她在最上面，整个冬天也得待在那里。地被冻得结结实实，就像一个大冰块，有一段时间气温甚至降到了零下。虽然积雪像一条厚厚的毯子盖在上面，可是，床还是冷冰冰的，怪不得一连好几个月欢欢都没有孵化。

When she did, poor little baby, she was buried alive under ground, and had a hard time of it pushing and pushing and pushing up; for she was the top of the brood and had to open up the way for her brothers and sisters as well as for herself.

等到她真的孵化出来，哦，可怜的小宝贝！她简直就是被活活地埋在了地底下，那段日子真是苦不堪言！她不得不使劲地往上挤呀挤，直到挤出地面。要知道，这窝卵里，她在最上面，所以，她得开路，为她的兄弟姐妹，也为她自己。

But the world she found when she crept out of that hole was worth working to get into, for it was spring-time and the sun shone and there were some warm stones near by; and so Baby Carol hopped about and ate whatever she wanted, and was, so far as anyone could see, quite happy.

不过，从那个洞里爬出来后，她来到了一个新世界。这个世界真是值得一来，以前的辛苦可没白费。因为春天已经来临，阳光灿烂普照着大地，附近还有一些热热乎乎的石头。于是，小宝宝欢欢蹦来蹦去，想吃什么吃什么，在大家看来，她十分快乐！

All that, of course, was long before she could fly; for a grasshopper never has any wings until she is grown up, though every time she moults her brown skin, the little pads, that will some day be fans and fan-covers, grow bigger and bigger.

Yes, that was all many weeks ago, and Carol was now grown up and old enough to have a mate.

当然，所有这一切都是陈年往事了，那个时候她还不会飞呢。因为，尽管蚱蜢每蜕掉一层褐色的皮，小翅芽就会长大一点儿，而这些小翅芽总有一天都会变成扇子和扇子罩，但是，在长大之前，蚱蜢向来是没有翅膀的。

没错，这都是好几个星期以前的事情了。现在，欢欢已经长大，都可以找对象了。

So we must not be surprised to see Carl Grasshopper talking pleasantly with her one day when they met on a large rocky place in the sun. But you may be surprised to know where he kept his voice, for though Carl had a throat and a mouth and lips, he used them to eat with and never spoke a word through them in his life. No, when Carl talked to Carol he did it by rubbing his great hind-legs against his sides; and

this seemed to her the most natural way in all the world to be spoken to, for it was the way her father had talked to her mother. It was a gentle scraping sort of sound, and you would have to be near to notice it. But Carol was sitting on the same stone, and her ears, which she kept in as queer a place as Carl did his voice, could no doubt hear him very well.

所以,如果有一天,当蚱蜢欢欢和蚱蜢土儿在一块开阔的石头地里相遇,顶着烈日起劲聊天的时候,我们一定不要大惊小怪。但是,你还是会奇怪,他的声音在哪里呢?因为,尽管土儿有嗓子,也有嘴巴和嘴唇,但是,那都是用来吃东西的,他可从来没有用它们说过一句话呀。的确没有。土儿需要和欢欢说话的时候,就用发达的后腿摩擦身体两侧。在欢欢看来,这样对她讲话是这个世界上最自然不过的事情了,因为她爸爸就是这样跟她妈妈讲话的。那是一种很轻的沙沙声,只有到跟前才能听得到。欢欢和土儿坐在一块石头上,虽然她的耳朵和土儿的一样,也不知道藏在哪个奇怪的地方,可是,毫无疑问,她能听到他的话。

Carl could sing, too, and his song was louder than his talking, and he had such a pretty way of doing it,

that poets, who love pretty things, have sometimes sung about Carl's song in their verses. That makes a sort of double song, doesn't it?

土儿还会唱歌呢，而且他的歌比他的话响亮多了。他唱得很美，有时候，连诗人都在诗里赞美他的歌呢，因为一切美好的事物都是诗人的所爱。土儿和诗人都在歌唱，简直就是一种二重唱，是不是？

When Carl wanted to sing, he would spring into the air about three feet, and hover almost in one place, with his wings spread and fluttering so fast that they made a clacking sound where the edge of the fans clicked against the edge of the covers.

想唱歌的时候，土儿就跳到大约三英尺高的地方，开始盘旋，其实，他几乎就是围着一个地方打转转。他张开翅膀使劲地拍打，拍得很快，扇子的边敲在罩子的边上，发出一种毕毕剥剥的声音。

And if poets have liked his song, is it strange if Carol was pleased?

所以，如果连诗人都喜欢他的歌，那么，欢欢喜欢他的歌也就不足为奇了。

But Carol could not spend all summer listening to music. She must hurry and get her eggs all buried before cold weather came. Carl did not help his mate dig a single hole; but you must not blame him for this, for his father had never dug a hole nor his grandfather nor his great grandfather nor his great-great-grandfather, nor any father as far back as you can count. So you see it was not the fashion in Carl's family to dig holes for their mates to lay their eggs in.

可是，欢欢可不能整个夏天都沉迷在美妙的歌声里。她必须加快速度，赶在天冷之前，把她的卵都埋起来。土儿是连一个洞都不会帮他老婆挖的。可是，你千万不要责怪他，要知道，他的爸爸从来也没有挖过一个洞，他的爸爸的爸爸、他的爸爸的爸爸的爸爸、他的爸爸的爸爸的爸爸的爸爸都没有挖过洞，你再往前数多少辈，都这样。所以，你就知道了，土儿家族是不兴为老婆挖洞埋卵的。

And, after all, Carol had plenty of time to do it

herself, and the tip of her body was all fitted up with the nicest little tool for boring into the ground—the four parts of it pushed down just right, and I think she really liked to use this little boring tool. And of course she really liked to put her precious eggs there in slanting rows in the soft froth that soon hardened about them, keeping them clean and safe all winter.

好在，欢欢的时间比较充裕，完全可以自己完成这项任务。她的尾部有一个世界上最棒的小钻土器。这个小钻土器的四个零件能恰到好处、准确无误地钻进地里。我觉得欢欢真的很喜欢用她的这个小钻子。当然了，她也真的很喜欢把自己的宝贝卵斜排着放到软软的泡沫里。泡沫很快变硬，把卵儿们包起来，好让他们干干净净、安安全全地过冬。

So her own queer little nest was left in the earth just as her mother's had been—that good old Earth that takes care of what is planted in her, —the same good old Earth to whom we owe, in many ways, our own life, just as surely as Carol's babies owe theirs to her: even those of us who are far from her, in a city flat way up in the air, instead of down by the country roadside, like a grasshopper in the sun.

因此，和她的妈妈一样，欢欢也把自己奇怪的小窝留给了大地，留给了慈祥的大地老妈妈，因为只要交代给她的东西，她都会悉心照料。其实，不仅欢欢宝宝们的生活多亏了慈祥的大地老妈妈，在许多方面，我们的生活也多亏了她，就连我们这些离她很远、高高在上居住在城市半空楼房里的人们，也和烈日下乡间小路边的蚱蜢一样，要感谢慈祥的大地老妈妈。

9

Ann Gusti's Circus
假死虫安妮的马戏表演

Ann Gusti was chewing a buttercup-leaf. If you did such a thing yourself, I dare say it would make your tongue smart a bit. But Ann Gusti belonged to a family of blister beetles, and a peppery salad tasted very good to her. In fact, she liked it better than any other food. For this reason some people name her Buttercup Beetle. Besides that, she is often called Oil Beetle, because she can drop oil out of her joints. And when she was a baby, just hatched from an egg, she was so funny that people said she was a Tri-un-gu-lin, and every time she changed her clothes somebody gave her a new name. So you see that by the time she was

grown up she had plenty of them; but I like Ann Gusti best of all.

假死虫安妮正嚼着一片毛茛叶。如果你也嚼一嚼，我敢肯定你的舌头会痛。不过，假死虫安妮是斑蝥家族的成员，对她来说，辣乎乎的色拉味道很不错。因为这个，有些人为她取名“毛茛甲虫”。另外，除了这个名字，她还经常被叫作“泌油甲虫”，因为她能从她的关节里往外滴油。她刚刚从卵里孵化出来还是个小宝宝的时候，样子非常好玩，所以，大家都说她是“三爪幼虫”。总之，她每换一次衣服，就有人给她起一个新名字。所以，你看，等她长大的时候，她就已经有很多名字了。可是，我最喜欢“假死虫安妮”这个名字。

Before she climbed up the buttercup-stem she put on the very last dress she was ever going to have, and a pretty one it was, too. It fitted her nicely, for, of course, like all beetle dresses, it grew on her body inside the dress she Wore before; and when it was ready for her to use, she crept out through a rip in the old one; and there she was, all spick-and-span!

她是换了一条连衣裙之后才爬到毛茛秆上的。这是她换的最后一条连衣裙，也是她仅有的一条连衣裙了，非常漂亮。当然，穿着也很合身，因为，像所有的甲虫连衣裙一样，这条连衣裙也是长在她身上的，就在她“穿”的那条连衣裙下面。等这条新衣裙备好能穿的时候，那条旧连衣裙就裂开一条缝，她就从里面爬出来了。所以，你瞧她现在，从头到脚，一尘不染、焕然一新！

It was a lovely dark-blue color, with some broad black stripes that were as shiny as satin. And on the back were two wing-covers that did very well for trimming, though they were not of much use, for there were no wings under them.

她的最后一条连衣裙是漂亮的深蓝色，上面还带着宽宽的、像缎子一样光滑鲜亮的黑色条纹。裙子背后还有两个翅膀罩，虽然，装饰效果不错，可是却没有什么实际用途，因为翅膀罩下面根本就没有翅膀。

No, poor Ann Gusti never had a ride with her own wings in her life. But perhaps we need not be sorry for her, because she had one good ride through the air for all that, one day when she was a baby and her name was Tri-un-gu-lin. As it was one of the

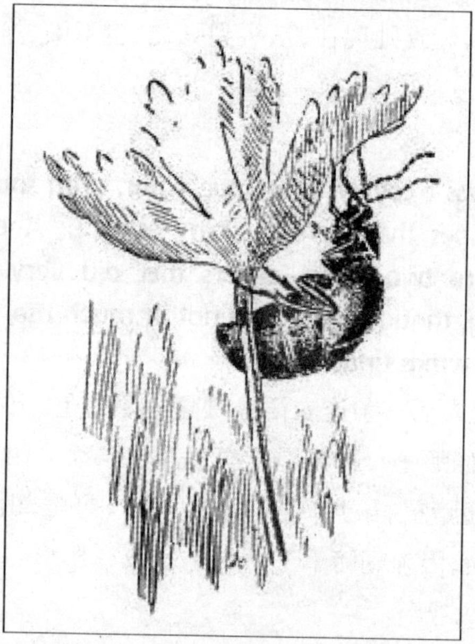

Before she climbed up the buttercup stem she put on her last dress.
她是换了一条连衣裙之后才爬到毛茛秆上的。

most wonderful rides in the world, that was enough to last her a lifetime.

的确没有。所以,可怜的假死虫安妮!她从未乘着翅膀兜过风。不过,我们大可不必为她感到难过,因为尽管如此,她小时候还是痛痛快快去天上兜过一次风的,那个时候她还是一个小宝宝,名字叫“三爪幼虫”。那次兜风举世无双,简直美妙极了,够她回味一辈子的!

This was the way it happened. When Ann Gusti hatched out of an egg, she was hungry. Of course she was. All babies are hungry, or else they wouldn't eat and grow up; and most insect babies can find something to eat very near at hand.

事情是这样的。假死虫安妮刚从卵里孵化出来的时候就饿了。她当然要饿了,因为所有的小宝宝都饿,如果不饿,他们就不会吃东西,不吃东西他们就长不大。大多数昆虫宝宝刚一出生就能找到吃的。

But Ann Gusti couldn't. That was before she learned to eat buttercup-leaves!

可是,假死虫安妮却不能。那是她还没学会吃毛茛叶时候的事情。

In fact, there was only one sort of food in the whole wide world that would agree with her when she was wee and wore her first baby clothes. Now that food was way, way off in a bee's nest, and she didn't know where the nest was. Most babies in a fix like that would have starved to death.

But Ann Gusti didn't! Oh, no! She just climbed up into a flower and waited.

其实,她只有不丁点儿大、还穿着第一套宝宝服的时候,全世界只有一种适合她吃的东西。可是,那东西在离她很远很远的一个蜂巢里,她也不知道那个蜂巢在哪里。大多数小宝宝如果身处这样的困境,肯定早就饿死了。

可是,假死虫安妮不会!噢,绝对不会!她爬到一朵花儿上,开始耐心等待。

By-and-by a bee came buzzing to the flower, and then, quick as a flash, wee Ann Gusti grabbed hold of the hairs on Mother Bee's leg; and that is how she got her ride and that is how she got her breakfast. For she

clung tightly to the hairs, and off she rode wherever Mother Bee went, from pretty flower to pretty flower, and at last right into the bee's nest, where she made herself at home and liked the food and stayed.

没过多久，一只蜜蜂嗡嗡嗡地飞了过来，落在了这朵花儿上。说时迟那时快，小不丁点儿假死虫安妮一下子抓住了“蜜蜂妈妈”腿上的茸毛。就这样，她兜了一次风；就这样，她吃上了早餐。她紧紧地抓住“蜜蜂妈妈”身上的茸毛，“蜜蜂妈妈”飞到哪儿，就把她带到哪儿。他们拜访了一朵又一朵漂亮的花儿，最后，正好落在了那个蜂巢里。到了那里，她一点儿也没感到拘束，还有吃的东西，她喜欢，于是她留了下来。

If you know a smarter baby than Ann Gusti, or a stranger ride to take before the first breakfast, I hope you 'll write a story about it.

如果你知道还有哪个宝宝比假死虫安妮还聪明，或者，如果你知道第一次早餐之前，还有比这个还奇特的兜风，那么，我希望你把它们编成一个故事写下来。

So with a ride like that to start life with, it is no

wonder that Ann Gusti grew up to be a clown in a circus.

假死虫安妮刚一出生就兜了这样的一次风，怪不得她长大后成了马戏团里的小丑演员。

Did you think that a circus had to be under a cloth tent, with tigers and lions in cages, and trained horses and dogs there to do tricks?

你以为马戏团必须得在布帐篷里表演吗？把老虎和狮子关在笼子里，让训练有素的马和狗表演各种把戏？

Well, Jack and Jane liked that kind, too; but they went to a different sort of circus with Uncle David one day as a special treat, and they both said his kind is the best there is.

是呀，杰克和简儿也都喜欢那样的马戏表演，但是，有一天，大卫叔叔特地款待他们，邀请他们去看了一场不一样的马戏表演。他俩都说大卫叔叔带他们看的马戏表演是世界上最好看的。

The top of the tent was the prettiest blue you can ever see, and at one side there were trees with the earliest ripe apples on them, and at another there was a river with a rocky bank and a great flat stone with a bonfire burning on it; and at a third side there was a meadow; but neither Jack nor Jane could tell what was on the fourth side, because Ann Gusti was playing her funny tricks in the meadow; and so that was as far as they got.

那场马戏表演的帐篷很不一般。篷顶碧蓝碧蓝的，你可能从来都没见过那样漂亮的蓝颜色。帐篷的第一面墙上是树，最先成熟的苹果早已挂满了枝头。帐篷的第二面墙是一条河，河岸岩石嶙峋，其中有一块石头又大又平坦，上面还有一堆熊熊燃烧的篝火。帐篷的第三面墙是一片青草地。帐篷第四面墙上有什么，杰克和简儿都说不上来，假死虫安妮正在青草地上表演她那有趣的把戏呢，所以，他们没顾上往那边看。

As I said to begin with, Ann Gusti was chewing a buttercup-leaf. She was right in plain sight at first, but the minute the toe of Uncle David's boot hit the buttercup-stem she was gone. Now you know she couldn't fly, because she had no wings under her

wing-covers; and if a bee had come along just then it would have done no good, for by this time she was grown up and much, much bigger than a bee; so she couldn't have ridden off that way, either.

一开始我就说过，假死虫安妮正嚼着一片毛莨叶。起初，她就在那里，一目了然。可是，大卫叔叔的靴子尖刚一碰到毛莨的茎秆，她就不见了。你知道的，她不会飞，因为她的翅膀罩下面根本就没有翅膀。那个时候，即使有蜜蜂过来也没什么用。因为这个时候，她早已经长大了，可比蜜蜂大多了，所以，她也不可能扒在蜜蜂腿上腾空而去。

After a hunt among the leaves she was found on the ground, lying on one side with her head held so close to her body that her yellow neck did not show. Her little feelers were reaching up, but they were still; and her legs looked limp, and there were oily drops coming out at the joints.

他们拨开叶子找了好半天，才在地上发现了她。只见她侧身躺在地上，头紧紧地缩在身子里，连她那黄灿灿的脖子都看不见了；一对小触角朝前伸着，可是，一点儿动静都没有；腿松松垮垮地耷拉着，关节里正往外滴油乎乎的东西。

"Oh! " said Jane, "she's dropped down dead! "

“噢！她摔死了！”简不由得喊道。

That shows how Ann Gusti fooled them! That shows how well she could do her trick ! No trained dog could have looked deader if he had practiced a year!

知道了吧？假死虫安妮多会骗人！知道了吧？她的把戏耍得有多高明！就算一条训练有素的狗训练一年，装起死来也赶不上她！

The children lay down on the ground with their chins in their hands, and waited and waited and waited. Just before their necks ached so that they couldn't wait another minute, Ann Gusti wiggled her toes. Then she moved her feelers the least little bit. Then after a while, as everything was quiet, she got up on her feet and climbed the buttercup-stem and went on chewing a leaf as if there was nothing at all the matter with her.

Ann Gusti wiggled her toes.
假死虫安妮的脚指头动弹了一下。

And there wasn't. She was just fooling them. That was Ann Gusti's trick.

孩子们双手托着下巴趴在地上，等呀等，一直等到脖子发酸，连一分钟都不能再等了。就在这个时候，假死虫安妮的脚指头动弹了一下。接着，她的触角也微微动了一下。又过了一会儿，等四周静悄悄的，没有一点声息的时候，她立起身来，爬到了毛茛秆上，好像没事人儿似的继续大嚼起毛茛叶来。的确没事。刚才她只不过是骗骗他们。那是她表演的把戏。

"She's a clown! " said Jack.

And that is how it happened that Ann Gusti had a circus on Labor Day before school began, when the friends of Jack and Jane could come.

“哇，她是小丑演员！”杰克喊道。

于是，就有了假死虫安妮的劳动节马戏表演。那时还没有开学，所以，杰克和简儿的朋友们都能来看。

The children played the woodshed as a tent. Beside the tent door there was a sign made by cutting big letters out of a newspaper and pasting them on

card-board to make the words:—

SIDE SHOW

THREE WONDERS OF THE WORLD

1. ANN GUSTI, THE FAMOUS FAT CLOWN

2. THE THINNEST LIVING SKELETON

3. THE SPRY AMERICAN ACROBAT

孩子们把柴火棚当成马戏帐篷。用报纸剪出大大的字母，拼成一个个单词贴在硬纸板上，当作告示牌，放在了帐篷的门旁边。

杂耍表演

世界三大奇观

1. 假死虫安妮，顶级“小丑人”

2. 顶级骷髅“芦柴棒”

3. 身手敏捷的美国杂技演员

Inside there were three cages for the animals. Ann Gusti was in the middle one, chewing buttercup-leaves. And when Jack touched her and said, "Now, you are a dead beetle," she would tumble down to the bottom of the cage and play she was dead—every single time. Good Ann Gusti—she knew her trick and did it, just as her mother and father and all the Gusti family had done as far back as any one can remember! They were

all clowns, those Gustis.

他们的马戏帐篷里面有三个关着马戏动物的笼子。假死虫安妮在中间的那个笼子里,她正嚼毛莨叶呢。只要杰克一碰她,说声“喂,你死定了,”她就滚到笼底,开始装死。假死虫安妮是个好演员,就像她的爸爸妈妈和那些能够数得着的假死虫先辈那样,她对自己要表演的节目了然于心,完成得很好。他们都是小丑演员,假死虫家族的所有成员都是。

The minute they hatched out of their eggs they were ready for their bee-riding trick; and then late in life, when they wore wing-covers that didn't cover any wings, they played they were dead before they were. In fact they had all saved their lives many a time by playing they were dead.

刚从卵里孵化出来的那一刻,他们就已经准备好了要玩跟着蜜蜂去兜风的把戏;后来,他们戴上了没有翅膀的翅膀罩,还没死呢,就玩起了装死的把戏。实际上,他们大家都通过装死才多次得以逃命。

So, being used to this trick, Ann Gusti could do it

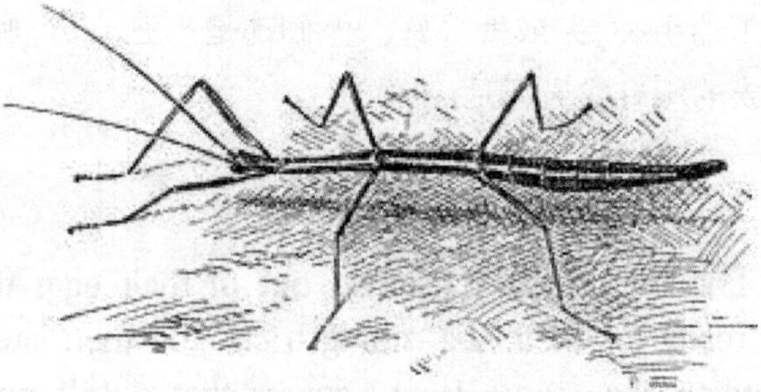

The thinnest living skeleton
顶级骷髅“芦柴棒”

just as well with everybody watching her as she could out in the field when some bird or little animal came too near.

所以，习惯成自然。在旷野里，鸟儿或小动物靠得太近时，她能成功完成表演；大家围着她看的时候，她照样也能成功完成表演。

In the cage to the right of Ann Gusti was the "Thinnest Living Skeleton," whose body looked like one little twig and her six thin legs like six others. She was a sort of distant cousin to Carol and Gryl, but you would never think it to look at her. They both had large thick strong jumping hind-legs to hop with, and if the "skeleton" had ever tried to hop, it would have made anybody laugh to see her. But I don't think she ever tried.

假死虫安妮右边的笼子里关的是“芦柴棒”。她的身子像一根小树枝，她的六条小细腿像六根小树枝。她和蚱蜢欢欢、蟋蟀灰灰有点亲戚关系，好像是远房堂姐妹或表姐妹什么的。可是，看她那样子，你永远也不会往这方面去想。欢欢和灰灰的后腿都粗壮发达，弹跳能力很强，可是，如果“芦柴棒”也能跳起来试一试的话，所有看到她的人都会忍俊不禁。不过，我想她从来就没试过。

As she had no wings or wing-covers whatever, she couldn't fly like Carol and she couldn't fiddle like Gryl.

由于她没有翅膀或翅膀罩之类的东西，她不会像蚱蜢欢欢那样飞，也不会像蟋蟀唧唧那样拨弄琴弦。

In fact, there wasn't much she could do but look like a twig. But she did that perfectly! If a person can do just one thing very well indeed, she gets along pretty well in this world. The "skeleton" had always been in a little oak tree before she came to the circus; and if you had tried to see her there, you would have found out that she could hide herself in plain sight on an oak-branch as well as Carol could hide herself in plain sight on the sand. Maybe that's how she came to look so like a twig that people call her a "walking-stick."

实际上，除了摆出一副芦柴棒的样子之外，她没什么会干的。不过，摆样子这件事她做得真是棒极了，简直是惟妙惟肖！一个确确实实有一技之长的人在这个世界上是能够应付自如的。“芦柴棒”来马戏团之前，一直住在一棵小橡树上。如果你以前去那里找过她，你就会发现她的隐身术和蚱蜢欢欢一样高明。蚱蜢欢欢能躲在沙土地里让你视而不见，而“芦柴棒”让你视而不见的高招就是躲在橡树上。也许，就因如此，她才长得这么像树枝，结果，人们都叫她“会走路的棍子”。

In the cage to the left of Ann Gusti was a beetle who had a queer spring on the under side of his body; and whenever he found himself on his back, he had the funniest way of getting on his feet you ever thought of. He would jerk his spring with a "click" sound, and that would throw him way up in the air, and then he would come down right-side up. If he didn't do it the first time trying, he would keep on clicking and jumping until he did.

So altogether Ann Gusti's side-show pleased the children and did the little animals no harm.

假死虫安妮左边的笼子里关的是一只小甲虫。他的身子下侧有一个奇怪的发条。只要他仰面朝天跌倒在地，他都会重新翻过身子站起来，那样子简直好玩极了，你连想都想不到。只见他“咔嗒”一声拉开那个小发条，一下子把自己抛到半空中，然后，腿脚朝下落到地

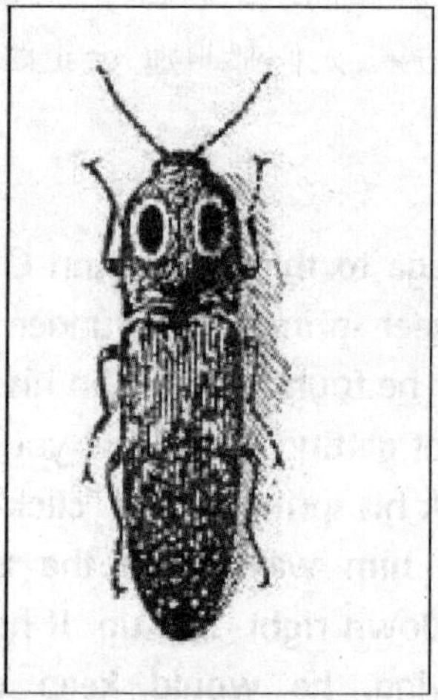

A beetle who had a queer spring
一只有奇怪发条的甲虫

上。如果第一次翻身没有成功，他就会不断地咔嗒、咔嗒拉发条、一次又一次地往上跳，直到把身子翻过来为止。

总之，孩子们很喜欢假死虫安妮的杂耍表演，再说，这些表演对小虫子本身没有什么伤害。

By the next day, though, she became tired of her cage and tried her best to get out. She climbed up to the top and clung with her six little feet at the crack, and pushed and poked and bumped with her head, trying to find a place where she could squeeze through. If Jack lifted the top, she would stop just where she was as if frightened, and keep still without moving a feeler or toe. When the cover was put down, she would wait a long time, and then begin all over again, creeping round and round and round the top of the cage, clinging at the crack with her feet, and pushing, pushing, pushing, with her hard little head.

可是，到了第二天，假死虫安妮就厌倦了笼子里的生活，她竭尽全力想爬出去。她爬到笼子顶上，六条腿紧紧地抓住笼缝，伸出头，连推带戳，还不停地撞，想找到一个地方挤出去。可是，如果杰克掀开笼顶，她就好像很害怕，立刻停下来，待在原地，触角和脚指头一动也不敢动。放下笼盖后，她还会一动不动地等很长时间，然后才重整旗鼓，又开始在笼顶上爬呀、爬呀，六条腿紧紧抓着笼缝，伸着她那坚硬的小脑袋，使劲地推了又推，挤了又挤。

That made both Jack and Jane so sorry for her that they gave her back to Uncle David, who took her home to her own wide meadow. And the last that was ever known of Ann Gusti, she had the great big beautiful sky for a circus-tent, and the little clown was chewing a buttercup-leaf whenever she felt hungry.

看她这样，杰克和简儿都难过极了。于是，他们把她还给了大卫叔叔。大卫叔叔把她送回了辽阔的青草地，让她回到了自己的家里。我们最后听到的关于假死虫安妮的消息，就是这位小巧可爱的小丑演员已经把广阔无垠的美丽天空当作了表演马戏的帐篷，而且，只要饿了，就大嚼毛茛叶。

10

Gryl, the Little Black Minstrel
蟋蟀唧唧，黑乎乎的吟游小歌手

A little bird with a blue back was going in and out of a round hole in an old tree near the river. Now and then he would stop and say "Tru-i-lee, tru-i-lee," very softly to his mate. He sang as if he liked the hole, and perhaps he did; for was this not the place where he and his mate had nested in the spring? And now, before they flew far off to the south, had he not come back again to sit near the old doorway and sing to her?

河边有一棵老树，树上有一个圆圆的洞。一只背上长着蓝色羽毛的小鸟从洞中进进出出，忙个不停。他时不时地停下来，呼唤他的另一半："快进来，快进来"，声音温柔极了。听声音,他好像很喜欢这个洞。也许的确如此。毕竟,春暖花开的时候,他和她曾经在这里筑巢做窝;现在,他又一次回来坐在熟悉的家门口,就是为了向她一展歌喉,然后,他们将离开这里,一起飞往遥远的南方。

One tree on the river-bank was red and one was yellow, for it was an October day and their leaves were no longer green. Under them the quiet water seemed red and yellow, too, as if a big looking-glass lay there for the trees to see themselves in.

河岸上有一棵红色的树,还有一棵黄色的树,因为金秋十月,树叶已经不绿了。树下,静静的河水躺在那里,红一块,黄一块的,像是为岸上的树准备的一面大镜子。

With a bluebird singing his good-bye song and the gay leaves making bright places in the water, it was a day to be happy in.

蓝色知更鸟的告别欢歌,还有缤纷的树叶倒映在水中的靓影!今天可真是个好日子呀!

There he sat, the happy little fellow.
他就坐在那儿,这个快乐的小家伙。

And Gryl Cricket was happy, even though he did not listen to the bird or look at the river. He was singing, too, the only song he knew. A funny sort of song it was, with all the words in the verse alike, and every verse just like the one before, and not much change in the tune. But although it was the same music over and over again, it had different sounds for all that. When near by, it was something like "Gr -gryl -1, grgryl-1, gr-gryl-1," tinkled oh, so quickly on a tiny bell; and when it was far off, it was more like "Cri-cri-cri-cri."

虽然蟋蟀唧唧没有侧耳聆听鸟儿的欢叫，也没有举目朝水里张望，可是，他很开心。他也在欢歌，他只会唱一首歌。这首歌有点可笑，每句话的词都一样，每句话都在重复上一句，曲调也没变化。可是，尽管整首曲子重复来重复去，没什么变化，听起来却不一样。近听有点像“嘀铃铃——嘀铃铃”，啊，清脆得就像是一个快速摇动着的小铃铛；远听更像是“唧唧——唧唧——唧唧——唧唧”。

Whether far or near, it was very good music for an October day. Anyway, Gryl liked to make it: you may be sure of that, for he kept at it, off and on, all the

warm sunny part of the day. And in the summer, before the nights were so cool, he sang in the dark, too.

无论近听还是远听，这歌声都是金秋十月的美妙享受。不管怎么说，唧唧喜欢唱这首歌，这一点可以肯定。白天，只要阳光明媚、天气暖和，他就会时不时地唱上一曲；炎炎夏日，夜晚也很热，即使天黑了，他也照唱不误。

With so much singing day after day, does his little black throat ache and his voice grow hoarse?

这样日复一日地唱，每天都唱个不停，他的小黑嗓子会疼吗？他的声音会哑吗？

Not a bit of it; for Gryl's music-box is not like yours and the birds'. It is not in his throat at all! He wears a fiddle on his back and sings, not through his throat, but with his wings.

一点儿都不会！唧唧有一个音乐盒，和你的音乐盒、鸟儿的音乐盒都不一样。他的音乐盒根本不在嗓子里！他背上背着一个小提琴，歌声是从翅膀那里飞出来的，不是从嗓子眼儿里唱出来的。

There he sat, the happy little fellow, before his open door, and fiddled. He lifted two wings and rubbed them together so that the row of tiny hard ridges on the under side of his right wing hit against the hard wrinkles on the upper side of his left wing, and that is the way he made music for himself.

瞧！这个快乐的小家伙，坐在敞开的家门前，起劲儿地拨弄着琴弦。只见他抬起两个翅膀相互摩擦，右翅下侧那一排又小又硬的隆起的棱，敲打着左翅上侧硬硬的褶皱。他就是这样自编自唱的。

His "Cri-cri-cri" just now seemed to be a song of thankfulness, as hungry people often say "grace" before they eat. Gryl had been too cold to get up for his breakfast, and now it was dinner-time and he was sitting before a feast and fiddling as he ate. "I have meat and I can eat," he seemed to be saying over and over to himself.

刚才，他那“嘀铃铃——嘀铃铃——嘀铃铃”的鸣叫似乎是在表达他的感激之情，就好像人饿了要吃饭时，常常先做祷告一样。唧唧太冷了，他连早餐都没有起来吃。这会儿，已经是晚餐时间了，他正对着一顿盛筵，一边吃一边拨弄着琴弦。“我有肉吃，我能吃。”他似乎在一遍又一遍地自言自语。

If he was glad, it was no wonder; he had found, when he went hunting, a nice fresh leg of grasshopper which a blackbird had dropped near his home. Now Gryl liked a grasshopper-leg as surely as you like a turkey's drumstick for Thanksgiving dinner. So he munched and munched all by himself, and fiddled as he ate.

如果他心情愉快，那没什么可奇怪的。因为，出去觅食的时候，他发现了一条新鲜美味的蚱蜢腿，是一只乌鸫鸟掉在他家附近的。要知道，蟋蟀唧唧喜欢吃蚱蜢腿，就好像感恩节的盛宴上，你喜欢吃火鸡腿一样。所以，他咯吱咯吱自顾自地大嚼特嚼，一边吃还一边拨弄琴弦。

Sometimes at parties, or in lunch-rooms, we have someone play to us during dinner. But we have to hire

our meal-time music. A man can not play a fiddle on his back and eat at the same time. Funny little Gryl could do just that, and it would make you laugh right out loud to see him do it.

我们在宴会上或者午餐室用餐的时候，有时候会有人为我们演奏音乐。可是，这种音乐得付钱。再说，没人能做到一边拨弄背上的琴弦，一边吃东西。可是，可爱的小唧唧却能做到，而且他边吃边拉琴的样子，一定会让你哈哈大笑。

The longer Gryl nibbled and fiddled, the more restless Taffy grew. Taffy Cricket, I forgot to say, was Gryl's neighbor, and he lived so near that he could smell Gryl's marrow-bone; and, like his namesake in the old verse, Taffy was a thief.

唧唧啃得时间越长，琴拉的时间就越长，塔夫就越焦躁不安。哦，忘了说了，蟋蟀塔夫是唧唧的邻居，住得很近，近得都能闻到唧唧啃出的骨髓味。就像老儿歌里的那个同名同姓的小偷一样，蟋蟀塔夫也是个小偷。

Perhaps if Taffy's home had been where the sun

warmed it earlier than it did Gryl's, Taffy would have wakened in time to find the bone first. But as it was, he had been too cold and lazy to start out until Gryl was half through his dinner. And the more he smelled that bone, the more he wanted it. So before long he crept out from under the slanting stone that covered his cave, and slipped very quietly in under the old piece of flat tin that made a roof for Gryl's dining-room.

如果塔夫住的地方能比唧唧住的地方先照到太阳，也许，塔夫就会及时醒来先发现这块骨头。可实际上，他太冷了，也太懒，所以，唧唧的晚餐都吃了一半了，他才出来。他越闻那块骨头越馋。不久，他就从斜盖在洞顶上的那块石头底下爬了出来，爬到那块旧罐头皮上，然后悄悄地溜下去，溜进了唧唧的餐厅。

While Gryl was making a merry, noisy tune over his meal, could Taffy come very softly and join the feast?

唧唧冲着他的美食大声欢唱的时候，塔夫能静静地过来和他共进晚餐吗？

Gryl and Taffy
唧唧和塔夫

He tried, and what happened to him was something like what would happen to you if you tried to take away a dog's bone. For wherever a dog has found his bone, he feels very, very sure, in his own mind, that it belongs to him, and no one except his master must go near him while he is eating. If we do, and get hurt, then it is our own fault and we must not blame the dog.

他尝试了一下，接下来发生的事情不难想象。试想，假如你要拿走一只狗正在啃的骨头会是什么结果呢？不管这个骨头是从哪里弄到的，狗儿们都会非常非常肯定地认为那块骨头是他的。狗儿啃骨头的时候，除了主人之外，任何人都不能靠近。如果我们非要靠近，受了伤，那也是我们自己的错，不能说狗的不是。

Perhaps Gryl felt that way, for he ran right up to Taffy and scared him off. The funny part of it was that he didn't stop fiddling while he ran! He just lifted his wings higher and fiddled louder, and this time he was not playing gently, "I have meat and I can eat," to himself: he was playing a threat to Taffy which meant, "Taffy Cricket this is my bone. I hunted for it. I found it. I brought it home. You lazy thing, you just go out

and hunt for your own dinner. Go away! Go away! Go away! "

也许，唧唧就是这么想的，他冲到塔夫跟前把他吓跑了。可笑的是，他冲过去的时候，也没忘记拨弄他的小提琴！他把翅膀抬得比平时高，琴声也拨得比平时大，不过这次，他可不是温柔地自言自语"我有肉吃，我能吃"，而是在威胁塔夫，"蟋蟀塔夫，这骨头是我的！是我千辛万苦找到的，是我带回家的。你这个懒东西，找你自己的晚餐去。走开！走开！快走开！"

Taffy must have understood him, even though what he said sounded like "cre-ek, cre-ek, cre-ek! " He understood so well that he did not even stop to turn round. He just stepped out backward and left Gryl, as he had found him, fiddling over his dinner.

唧唧的话听起来就像"唧啾——唧啾——唧啾"，但是塔夫肯定听懂了。他太明白这琴声的意思了，所以连身也没敢转就退了回来。不过，他只是退到了一旁，远远地看着唧唧一如既往地边吃边拉琴。

When Gryl had eaten plenty of food, he washed

one of his front feet by putting it into his mouth. Then he pulled one of his long feelers down with his foot, until he could wash that, too. After that he brushed his shiny black sides with his hind-legs; for this was Gryl's wedding-day and of course he must be clean as well as happy.

唧唧吃够了,就把一只前脚伸进嘴里洗刷干净。接着,伸出一只脚把一根长长的触角拉下来,一直拉到嘴边,也洗干净。最后,他伸出后腿把他那黑黝黝、闪闪发亮的身体两侧都刷干净。今天是唧唧举行结婚典礼的日子,他不仅要高高兴兴,而且还要干干净净。

In the afternoon he went for a walk in the sunshine. He was merry as could be, except when a man happened to come that way and shake the ground with his heavy foot. Then Gryl stopped fiddling and sat so still that even his feelers did not move. He stayed still a long, long time, and then crept off under some fallen leaves, to hide until the steps passed far away. He could never fly even when danger was near. His front wings were a fiddle and his hind ones weak, little folded things that were too feeble to carry him an inch. But he knew how to slip away and hide, as you would

find if you tried to catch him for a pet.

下午，他顶着烈日去散步。一路上，他甭提有多美了。可是，途中，有个人碰巧从那边走过来，重重的脚步让地面都微微颤动。唧唧立即停住脚，静静地坐在那里，连触角都不敢动一下。他一动不动地待了很长很长时间，然后，才爬到一些落叶下面藏了起来，直到脚步声渐渐消失在远处。即使危在眉睫，唧唧也绝对不会飞走。他的一对前翅是小提琴，而后翅折叠在一起，弱不禁风，根本无法带他起飞，哪怕一英寸都不行。可是，他知道怎么溜到一边躲起来。假如你试图逮住他当宠物，那么，你就会深有体会。

Toward night he played a soft tune to Lucy Cricket, who lived under a bit of bark near his own little dug-out.

夜幕快要降临了，他为蟋蟀露西拉起了轻柔舒缓的曲子。她就住在离自己的“防空壕”不远处的一小块树皮底下。

Now Lucy was black as a piece of coal, and her head was so bald and shiny that not a hair could be

seen on it anywhere, and she was dumb.

瞧,露西黑乎乎的,就像是一块煤炭。光秃秃的脑袋油光闪亮,连一根头发都找不到。另外,她还是个哑巴。

But she was not deaf, and she could hear Gryl's song and liked it. And what do you suppose she did? Do you think she painted her little black cheeks and put on a wig? Oh, no; she was an honest cricket and went out to her mate just as she was. As he was black as a piece of coal, too, and just as bald as she was, they really looked very much alike, except that her wings had no fiddles, and he had no long slender tail, but only two tail-feathers.

可是,她不聋。她能听见唧唧唱的歌,而且很喜欢。你以为她会怎么办?在她那小小的黑脸蛋上涂脂抹粉?带上假发套?噢,不会的。她是一只诚实的蟋蟀,所以,她就素面朝天地去见她的另一半。蟋蟀唧唧黑得也像一块煤炭,而且和她一样,脑袋也是秃秃的,他俩长得还真般配。不过,露西的翅膀上没有小提琴,而唧唧没有长长的细尾巴,只有两个尾羽。

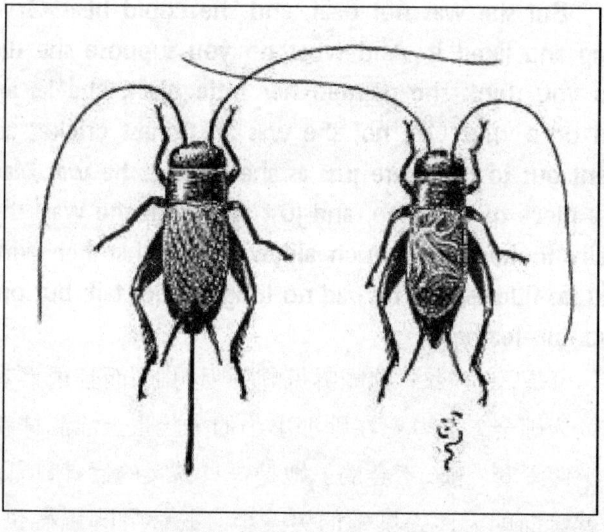

Lucy and Gryl Cricket
露西和蟋蟀唧唧

Well, Gryl went right on fiddling, and this time it was their wedding march. For a wedding supper they had a wild-lettuce salad, with some apple-sauce for dessert; for there was a wild apple tree on the bank, and the apples that had dropped down were nice and soft.

好了，唧唧不停地拨弄着琴弦，这次他拉的是他们的婚礼进行曲。婚宴上，他们吃野莴苣做的蔬菜色拉，甜点是苹果酱。河岸上有一棵野苹果树，从树上掉下来的苹果软软的，很好吃。

Then Lucy went back to her cave under the bark and Gryl found his way to his own little dug-out.

婚礼结束，露西回到树皮底下自己的洞里，而唧唧也一路摸回了自己的小“防空壕”。

After all, he did not need any one to help him about his meals. He rather liked going out to hunt for them himself. And he needed no one to show him how to finish his dug-out. He knew just how to pick up the hard little bit of dirt or stone in his mouth and carry it off out of the way, and just how to scratch out the soft places with his head and feet.

要知道，唧唧不需要谁来帮他准备一日三餐，他很喜欢亲自出去觅食。他也不需要谁来告诉他怎么完成他的“防空壕”。他知道怎么叼起硬硬的土块或石块，把它搬到一边。他也知道怎么头脚并用把虚软的地方刮出来。

It was well that he felt that way about it, for Lucy was far too busy with her own task to help Gryl or any one else. It took her all her time for a while to take care of her eggs. For these must be put down so that they would keep all winter.

幸亏他有这样的生活理念。因为，露西自己的事情都还忙不过来呢，哪里还顾得上去帮唧唧或者别的什么人？有一阵子，她所有的时间都用来照顾她的卵了。她得把这些卵安置妥当，好让他们安然过冬。

What better could she do with them than trust them to good old Mother Earth, who broods over the tiniest seeds that are left in her care? And Lucy's eggs were something like wee seeds. She planted them in the earth with the long slender back part of her body,

that looked like a tail but was really a tool to put eggs into the ground with.

除了把他们交代给慈祥的大地老妈妈之外,她还有什么更好的办法呢？哪怕是最小最小的种子,只要交给大地老妈妈照顾,她都会呵护备至。露西的卵就有点像小小的种子。她身体的后半部分又细又长，像个尾巴。实际上,那是一个工具,露西正是用这个工具,把她的卵一个个地种到了土里。

So while Lucy tended her eggs like a good little mother, Gryl finished his dug-out, making it deeper and deeper, until the days grew very cold and he fell asleep in his bedroom in the same earth that sheltered Lucy's eggs.

当小露西像一个好妈妈那样悉心照料着她的卵的时候,唧唧的"防空壕"也完工了。他越挖越深,天气也越来越冷,最后,实在是太冷了,他就睡着了。他的卧室也在大地的怀抱中,就是那个为露西的卵遮风挡寒的大地。

One day, a long time after that, the trees, which

had dropped their red and yellow leaves on the river-bank, unfolded some very fresh green ones. A bird with a blue back, who had flown far, far to the south, and north again, was hopping in and out of the hole in the old tree near by and whispering "Tru-i-lee, tru-i-lee" sweetly to his mate.

很久以后的一天，曾经将红树叶和黄树叶洒满河岸的那些树又吐出了新绿。附近那棵老树的树洞里，一只背上长着蓝色羽毛的鸟在进进出出，温柔甜蜜地低声呼唤着他的另一半："快进来，快进来"。他越过千山万水，飞到了遥远的南方，现在又飞回了北方。

It was spring, and time for Gryl to waken from his long nap and begin fiddling again. For there are always some crickets to welcome the spring—their songs do not belong just to the summer and fall.

这是春暖花开的季节，是唧唧从长长的睡梦中醒来，又一次拨弄琴弦的季节。春姑娘来了，总会有一些蟋蟀出来表示欢迎——他们的歌可不只是献给夏哥哥和秋大叔的。

As Gryl sat and fiddled before his open door, his little sons and daughters hatched in their egg-shells and climbed up out of the ground.

当唧唧坐在敞开的家门前拨弄着琴弦的时候，他那些小巧可爱的儿女们也孵化好了，纷纷挤破卵壳，爬出了地面。

They were wee midgets who looked much like old Daddy Gryl, with their bald black heads and hopping hind-legs. For a cricket is a cricket all the days of his life except while he is an egg. He isn't one thing first and then another afterward, like a caterpillar turning into a butterfly.

他们很小很小，简直就是老爸唧唧的袖珍版。一个个黑乎乎的脑袋秃秃的，后腿也能跳。除了卵阶段之外，蟋蟀一辈子都是蟋蟀。他绝不会像毛毛虫变蝴蝶那样，一会儿是这副模样，一会儿又是那副模样。

All a small cricket has to change to is a larger cricket. So there was not a great deal of difference between Gryl and one of his little sons except size and wings. For a baby cricket is like all other baby insects,

who must eat and grow up before they can have any wings.

小蟋蟀发生的唯一变化就是变成一只大蟋蟀。比如,唧唧的一个儿子,除了比唧唧小、没有长翅膀之外,看上去和唧唧没什么大的差别。蟋蟀宝宝和其他所有的昆虫宝宝一样,都必须先吃东西长大,然后才能长翅膀。

While he is growing, he must throw off his little black skin when it gets too tight, and from time to time his wing-pads will get bigger. Then, when he changes his skin the last time, there will be wings on his back instead of pads—not wings to fly with, but to fiddle with.

发育过程中,他必须时不时地扔掉小得不能再穿的黑色小外壳。每换一次外壳,他的翅芽就变大一些。后来,当他最后一次换掉外壳的时候,背上的翅芽就不见了,变成了翅膀。不过,这翅膀可不是用来飞的,而是用来拉琴的。

So, long before the bluebird will come to say

good-bye to his nest next October, Daddy Gryl's sons will be fiddling near their open doors, merrily, oh merrily, as crickets should.

金秋十月,蓝色的知更鸟又要来说再见了。可是,还没到那个时候,老爸唧唧的儿子们就早早地坐在敞开的家门口，欢欢喜喜、快快乐乐地拨弄起了他们的琴弦。这是蟋蟀的本分。

11

Luna's Thanksgiving
月形天蚕蛾卢娜的感恩节

The hazelnut bushes had dropped their leaves. They were dry brown leaves now, that rustled when the boys and girls waded through them; and when the wind lifted them up and made them dance, they flew here and there as if they were merry.

棒树丛里的榛子树开始掉叶子了。落叶已经枯黄，孩子们从上面走过时，它们就沙沙作响；风吹过时，它们就翩翩起舞，四处飘荡，似乎很高兴。

But there was one leaf that did not dance with the

others. It stayed still, for it was wrapped tightly about Luna's little silk-room. Luna had papered the outside of her room with the leaf before she spun the wall too thick to reach through. Now you can tell from that that Luna was a caterpillar; because what else could it be that would spin a room of silk?

可是,有一片叶子却没有加入跳舞者的行列。它静静地待着，原来，它被紧紧地包在了卢娜的小丝屋外面。卢娜是先用那片叶子把她的小屋包起来,然后才吐丝砌墙的。她的墙很厚,根本无法穿透。现在你知道了吧？卢娜是一只毛毛虫,会吐丝筑屋的还能是别的什么东西吗?

Yes, Luna had been a caterpillar, once upon a time in her spinning days—a big one, too, and a pretty one, a very, very pretty one. But after she had finished making her little home that had just one room in it, she rested a while and then she stopped being a caterpillar. It happened this way. Her last caterpillar clothes ripped down a back seam just behind her head, and her caterpillar skull cracked open like a nut. And the pale little body that lay inside was not a caterpillar any more. No, Luna was a pupa now, and grew darker and

darker until she was brown all over. There was nothing for her to do but to sleep, for her one room was a bedroom, and she had locked herself in until spring.

不错，卢娜曾经是只毛毛虫。而且，在吐丝的那段日子里，她还是一只大毛毛虫呢，一只漂亮的毛毛虫，非常非常漂亮。可是，等建完她那只有一间屋子的房子，并稍事休息之后，她就不再是一只毛毛虫了。事情是这样的：她换上的最后那套毛毛虫服沿着后脑勺的背中线裂开了一条缝，她的毛毛虫脑袋也像坚果一样裂开了。那个躺在里面的苍白小身体已经不是毛毛虫了；不是了！卢娜现在已经变成了一只蛹，身体的颜色也越来越深，到最后，整个儿都变成了褐色。除了睡觉，她无事可做，她唯一的房间就是卧室，她已经把自己锁在了里面，不到春天无法出来。

There was no pantry in her little house, and she could not eat. Not one crumb of food and not one drop of drink was she to have all winter. But she was so plump that we need not feel sorry for her, for she had spent all her life in a pantry until she made her bedroom, and had eaten enough to last her a long, long time. Her pantry had been an oak tree, and the shelves

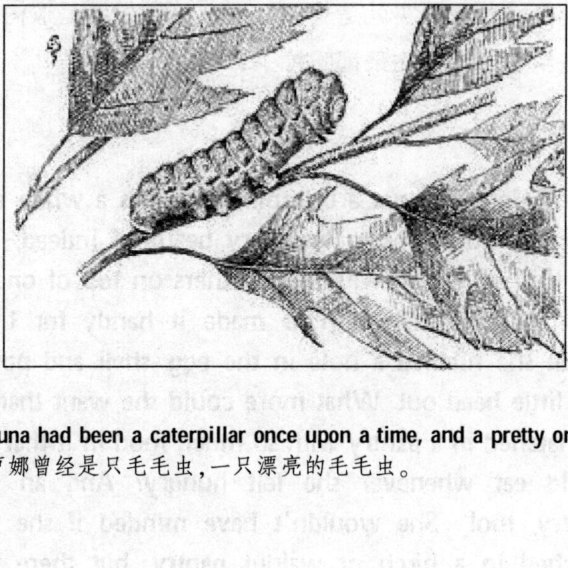

Luna had been a caterpillar once upon a time, and a pretty one.

卢娜曾经是只毛毛虫，一只漂亮的毛毛虫。

were the oak -leaves, and she ate up some of the shelves!

她的小房间里没有食品储藏室,所以,她吃不到东西。整个冬天,她连一点儿食物渣子都吃不到,连一滴喝的都没有。可她竟然长得胖嘟嘟的,我们根本不必替她难过。卧室完工之前,她一直都住在一个食品储藏室里,吃了很多的东西,足够她过很长很长时间的了。她的食品储藏室其实就是一棵橡树,储物架就是橡树叶,她甚至连一些储物架都吃掉了!

Before she was a caterpillar she was a white egg; and her mother, who was very beautiful indeed, had put this white egg with some others on top of one of the oak -leaf shelves. This made it handy for Luna when she nibbled a hole in the egg -shell and poked her little head out. What more could she want than to be hatched in a pantry with so much food in it that she could eat whenever she felt hungry? And an oak pantry, too! She wouldn't have minded if she had hatched in a birch or walnut pantry, but there was nothing she liked better than oak. Indeed, she liked an oak -leaf as well as a blue jay or a squirrel likes an acorn—maybe even better.

还没变成毛毛虫之前，她是一颗白色的卵。她漂亮的妈妈把这颗白色的卵和别的卵一起放在了一片橡树叶储物架上。这样，当卢娜咬破卵壳，探出小脑袋的时候，马上就可以有吃的了。还有比这个更令她心驰神往的吗？在食品储藏室里出生，到处都是食物，只要饿了就有吃的。再说了，这还是个橡树储藏室呀！当然，如果是在白桦树或者胡桃树储藏室里出生，她也不会介意的。可是，她最喜欢的还是橡树。的确，她喜欢吃橡树叶，就像蓝松鸦或者小松鼠喜欢吃松子一样，也许更胜一筹呢。

So of course she ate and ate, and I think she would have eaten for sixty days without stopping, except for one thing. That one thing was her skin. It was a funny sort of skin to have. It could stretch and grow for about a week, and then it would stop. Whenever Luna's skin stopped stretching she had to stop eating. As soon as this happened, she would spin a thin silk mat all spread out on the leaf, and tangle the little hooks on her ten fat clinging feet in it, and wait. While she waited she would lift her head and put her six thin front-legs together, so that she looked as if she were asking for something.

所以，她理所当然地吃呀、吃呀，我想，要是没有那件事情的话，她会一连不停地吃上六十天。那件事情就是她的外套。她的外套很好玩，能越撑越大，这个过程要持续一个星期。只要卢娜的外套不能再往大撑了，她就得停止吃东西。每当此时，她就吐丝织一个薄薄的丝垫，平铺到叶面上，然后伸出十个肥嘟嘟的抓脚放到丝垫上，把抓脚上的小钩子塞进丝垫里缠住，开始等待。这期间，她会仰起头，把六条细细的前腿并到一起，看上去就好像是在祈求什么似的。

What she needed was a new skin that would stretch farther, and a new head so that she could eat bigger slices of good oak-leaves.

她要一件会撑大的新外套，还有一颗新脑袋，好让她能吃下大块鲜美的橡树叶。

And sure enough she was going to have them both! For there behind her head that was too little for her, there was a place where her neck looked swollen. That was because a new head was growing inside her old tight skin; and as it grew, it pushed the little old

skull off until it looked something like the nose-bags men put on their horses when they feed them away from home, where there is no food-box.

当然了,这两样她都会有的！她外套后脑勺那儿太小了,脖子那儿有个地方都鼓了起来,那是因为在她紧绷绷的旧外套里,又长了一个新脑袋。新脑袋越长越大,最后就把原来的旧脑壳给掀开了。结果,她的外套看起来就像是远离家园的人们喂马时,由于没有马食槽而不得不挂在马脖子上的那种马料袋。

And at last her bigger new head inside her skin pushed the little old skull so far that rip, rip went the skin right round her collar, and off dropped the little old skull and out popped her new head! Then she crept out of the old tight skin through the collar-hole; and there she was as good as new, with a new suit of skin that would stretch for about a week. So she could eat another good long meal.

外套里面的那个新脑袋太大了,不断地往外挤,挤呀挤,最后竟然哧啦啦一声,领口周围那一圈整个儿裂开了,原先的那个小脑壳竟然掉了下去,新脑袋一下子就冒了出来！接着,她从领口爬了出来,把那身紧绷绷的旧外套甩在了身后。瞧！她焕然一新，刚刚换上的新外套足以应付大约一个星期，她又可以不慌不忙地吃个够了。

That was the way she ate and grew, and that was the way she changed her skin-dress. And every dress she had was a little different from the one she had before. They were all pretty dresses, and all green ones that did not show much when she was clinging to her green pantry shelves. The tips of her legs were reddish brown, and so was her mouth. She had some tiny red dots and some yellow ones on her dress, and some blue ones. There was a yellow line along each side and some in other places. These colors made her all the prettier, but they did not show very far away, and she really looked nearly as green as the leaves she ate.

她就这样吃了长、长了吃，就这样蜕皮换外套。她换的每一身外套，和原来的那身旧的比起来，都只有一点点不一样。所有的外套都漂亮，所有的外套都是绿的，让她爬在绿色食品储藏架上时，不太显眼。她的腿梢是红棕色的，嘴巴也是。外套上点缀着小红点、小黄点，还有小蓝点。她的身体两侧各有一条黄线，身上别的地方也有一些黄线。这些颜色让她越发漂亮。然而，

The moonbeams could find nothing lovelier than Luna.
月光找不到什么东西能与卢娜相媲美。

在远处是看不到这些颜色的。所以,看上去,她真的就和她吃的叶子一样绿。

When she had on her last suit of skin, she grew to be three inches long; and when she was about sixty days old, she left the oak pantry for the first time in her life. And she was never going back. Plenty of good fresh leaves lay all about her on every side; but greedy as she had been all her growing days, she seemed to know when she had had enough and did not stop to take another bite. She walked along the branch until she came to the trunk of the tree, and then down she crept head first—way to the ground. Then she wandered off for her very first walk on the ground.

换上最后一身外套的时候，她身长已经三英寸了。长到六十天的时候，她生平第一次离开了橡树储藏室，而且从此一去不复返。她的四周到处都是优质的新鲜树叶,尽管在发育旺盛的那段日子里,她很贪吃,可是,她似乎知道适可而止,所以,她现在并没有停下脚步再多吃一口。她沿着树枝走呀走,一直走到树干上,头冲下沿树干爬到了树下。然后,她离开树干,开始了平生第一次地面之旅。

It was her very last one, too; for by the time she had reached a hazelnut bush she felt like spinning. It was more than a thin mat to rest on, while she changed her dress, that she felt like making. There was a leaf near by and she felt like spinning this about her with silk and making her cocoon inside it. Cocoon is the name of her bedroom.

这也是她平生最后一次地面之旅。当她走到一个榛树丛跟前的时候,忍不住想吐丝了。这次她想做的可不仅仅是一个换衣服时休息用的薄垫子。旁边有一片叶子,她想在这片叶子上吐丝把自己围起来,然后在里面做茧。茧就是她卧室的名字。

The silk she used dripped out of a tube near her under-lip. It was wet and sticky at first, but as soon as the air touched it, it hardened into a fine thread. She spun these threads in little loops by swinging her head from side to side, and it took more loops than you can count to make enough to cover her all up, though the more she spun the shorter she grew. She seemed to shrink as she worked; and by the time the cocoon was done, it did not have to be more than an inch and a half to hold her.

她用的丝是从下嘴唇旁边的一根软管里滴出来的。刚滴出来的丝又湿又粘，可一接触空气，就变硬了，成了细细的丝线。她一边吐丝，一边摇头晃脑，从这边晃到那边，吐出的丝就绕在了她的身上。尽管她多吐一圈丝，身体就会缩短一点，可要把她全身都绕起来，仍然需要吐很多很多圈丝，多得数都数不清。她吐丝的时候，身体似乎也越来越短，所以，绕成的茧，长度不足一英寸半就可以把她裹起来。

That is how she came to be all snugly tucked up in a silk cocoon with a leaf wrapped about it. Then she changed to a little pupa inside, as brown as the hazelnuts the children were hunting.

她就这样把自己舒舒服服地裹在了丝茧里面，茧的外面还包着一片树叶。接下来，她在里面变成了一只小小的蛹，棕褐色，跟孩子们漫山遍野去捡拾的榛子颜色一样。

For it was Thanksgiving time, and Jack and Jane were nutting in Uncle David's woods. But Luna did not hear them as they laughed and shouted and waded

through the dry leaves, any more than she heard the big black birds calling "Caw! caw! " or the other big bird with a jacket almost as blue as the sky, scream at the top of his lungs that his name was "Jay! Ja-ay! " The woods might ring with noises, but Luna slept on. Boys and girls might have turkey baked brown and cranberries cooked to a beautiful red jelly and candy with hazelnuts in it for their Thanksgiving dinner. What did Luna care about that? She had eaten her fill long ago. The only Thanksgiving she needed was to sleep safely in her cocoon. That is how she spent the day. Alone and asleep! Sometimes she wiggled and turned in her bed, but she did not waken. She would wait until Spring called her, and then her long, long night would be over.

那天是感恩节，杰克和简儿正在大卫叔叔的林子里采坚果。他们脚踩干枯的树叶沙沙作响，一边走还一边高声欢笑，可是，卢娜却没有听见。咳，那些大黑鸟也在“呱！呱！”地大叫，她不也没有听见吗？还有，那个穿着几乎和天一样蓝的夹克衫的大鸟也在拼命大喊：“哇！哇—哇！”，她不也照样没有听见吗？随便林子里怎么吵、怎么闹，卢娜都照睡不误。就算在感恩节大餐上，无论男孩还是女孩都可以吃到烤得黄灿灿的火鸡，还有用蔓越莓做的红红的果酱，还有榛子馅的糖果，卢娜也不在乎。她早就吃得饱饱的了。感恩节那天,只要能在自己做的茧里安安稳稳地呼呼大睡,她就感激不尽了。她的感恩节就是这样过的:孤零零地呼呼大睡!有时候,她在床上辗转翻身,但是不会醒来。她要一直睡到春姑娘来叫她。那个时候,她的漫漫长夜才会结束。

Then she would break her brown pupa case and wet her cocoon with something that would soften the hard silk so that she could break her way through, and out she would come—out of her bedroom into the open world of sunshine and moonlight!

那时,她会弄破棕褐色的蛹壳,用一种能把硬邦邦的丝线变软的东西将身上的茧打湿,然后破茧而出。她会离开卧室,来到一个白天阳光明媚、夜晚月色皎洁的开阔世界!

At first her wings would be tiny limp flaps, and they would grow bigger and bigger and bigger until it would take a ruler five inches long to reach across them when they were spread. And they would grow so fast you could see them do it!

She would fly in the moonlight.
她会飞到月光里去。

刚出来的时候,她的翅膀是很小的薄片,软塌塌的,一点劲儿也没有。不过,它们会越来越大,最后大到展开的时候,要用一把五英寸长的尺子才能从这边量到那边。她的翅膀会在很短的时间内变大,所以,你都能看到它们变大的整个过程。

Their color would be the loveliest pale green there is in the world. Near the middle of each one there would be a tiny clear place, like a wee window with a pretty frame of white, black, and red. Across the front edge of the front wings would be a border of purple, and the hind wings would each end in a long part like a tail.

卢娜的翅膀会是淡绿色的, 世界上最漂亮的淡绿。每片翅膀靠近中间的地方都会有一小块是透明的,就像是一个袖珍窗户,窗框很漂亮,有白、黑、红三色。两片前翅的前部边缘会横镶一条紫色的边,两片后翅的末端会长得长长的,像两条尾巴。

Luna's body lying between the wings would be covered by beautiful white fluffy scales, the whitest

white that ever was seen—whiter even than the feathers on a white dove! Upon her head would wave two pale brown plumes. There she would be next summer—a moth so lovely that you would not want to touch her, but just look and look and look.

卢娜的身体位于翅膀之间，盖着茸茸的白色鳞毛，会非常漂亮。那白色是世界上最白的，甚至比白鸽身上的羽毛还要白！她的脑袋上会飘起两根浅褐色的羽毛。总之，明年夏天，她就变成一只美丽的飞蛾了。她会美得让你连摸都舍不得摸一下。你会只顾着看呀看，却怎么都看不够！

What would she do, this wonderful green and white and purple moth who had been a plump, brown sleeping pupa before that, and a pretty, growing caterpillar, green as an oak－leaf, before that, and a white egg first of all?

她，这只美妙绝伦的绿、白、紫三色飞蛾，将会做什么呢？变成飞蛾之前，她是一只胖嘟嘟的褐色睡蛹；变成睡蛹之前，她是一只漂亮的、越变越大的、和橡树叶一般绿的毛毛虫；而最开始，变成毛毛虫之前，她是一颗白色的卵。

She would fly in the woodland, but not by day. She was a moon moth and would wait for night to come.

她会在林地里飞,但不是在白天。她是一只月形天蚕蛾,她要等待夜晚的降临。

By night she would fly, and where, do you think? To the flowers with deep cups, and uncoil her long tongue to sip up the sweet drops at the bottom? Oh, no. That is what many moths would do, but Luna had no tongue and could not drink. That is the way Lunas are made. They eat, when they are caterpillars, enough to last them while they sleep the sleep of change. All moths are like that. But Luna's caterpillar food lasts her all her life, though many other moths sip from flowers at night as butterflies do by day.

她只在夜间飞,那么你想她会飞到哪里去呢?飞到花杯很深的花儿那儿去,伸出长长的舌头去把杯底的蜜露吸个一干二净?噢,不!许多飞蛾会那样做,可是,卢娜没有舌头,她不能喝东西。所有的卢娜们天生都这样。她们还是毛毛虫的时候,就吃很多食物,为她们的变身之眠做好充足的储备,所有的飞蛾都这样。可是,卢娜毛毛虫时候吃的食物能够维持她一生,而其他许多飞蛾夜晚从花儿那里吸吮营养，就好像蝴蝶白天吸吮花蜜一样。

No, Luna would not visit the blossoms that hold up their sweet-smelling cups. She would fly—have you guessed where?

不会。虽然树上的花儿纷纷举起了芳香四溢的花杯，可是，卢娜不会去她们那里做客。她会飞到——你猜出来了吗？她会飞到哪里去呢？

Yes, she would fly in the moonlight.

是的，她会飞到月光里去。

She would be beautiful enough to be a fairy, with her snowy robe and her pale green wings. You could call her a fairy princess if you liked, and say that her name is Princess Luna. And the prince would come to fly near her in the moonlight. Be sure of that—Prince Luna would be there, too. He would be dressed like the princess, with a white robe and fair green wings, and the plumes on his head would be even larger than

Would they call the princess, do you think? Or was there a fragrance in the whispering leaves that drew her to them? In some strange way the oak would tell her to come, and she would fly back to the tree she left so many months before. There upon the leaves she would put her eggs, her most precious gift to the world; and they would gleam in the moonlight like fairy pearls.

你想他们会召唤公主吗？或者，是窃窃私语的橡树叶有一种香味吸引了她？总之，不知用了什么奇怪的方法，橡树总会吩咐她过去，而她则会听话地飞过去，回到好几个月之前离开的那棵树上，而且就在那棵树的叶子上，她会产下她的卵。那是她送给这个世界最珍贵的礼物。这些小礼物会像童话中的珍珠那样在月光下熠熠闪光。

She had been like one of them herself a year before. And if, in the year to follow, there are to be Prince and Princess Lunas floating among the moonbeams, there must be pearly eggs upon the oak-leaves, just as later there must be cocoons wrapped in fallen leaves, each with a pupa in it brown as a nut at Thanksgiving time.

一年前，她自己曾经就像是其中的一颗卵。所以，来年，如果有卢娜王子和卢娜公主在月光中飘游，那么橡树叶上肯定会有珍珠般的卵，就好像再晚些时候，肯定会有包着落叶的茧一样，那些茧里面都有一个蛹，颜色和感恩节时候的坚果一样，是棕色的。

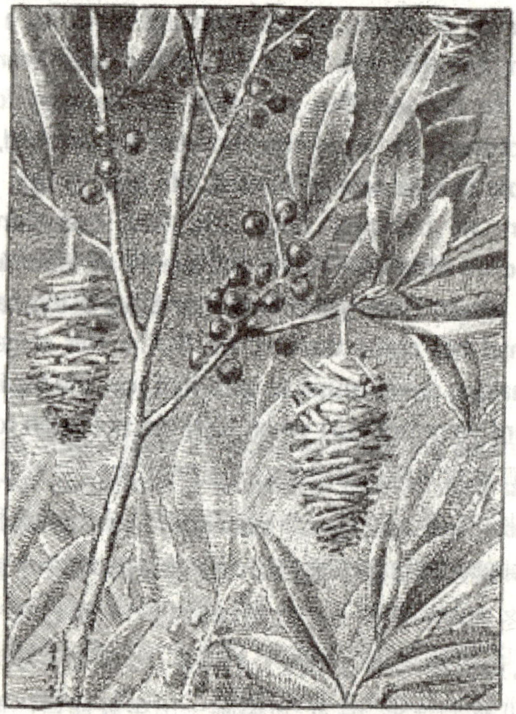

Keti lived all alone in a tiny log cabin.
凯弟孤苦伶仃地住在一个小木屋里。

心,即使凯弟对他的歌声充耳不闻,他也不会伤心。

Keti was so busy getting his meals and building his log cabin that little else mattered to him, just so he was let alone as his father and mother had left him. It seems strange to think of that little chap finding food for himself from the time he was a baby less than a day old and ready for his first bite. Yet very soon after he had poked his head out of the egg-shell, he had crept off and fed himself on a nice crisp holly-leaf salad. It agreed with him better than malted milk would have done, which you may be very sure was the reason Mrs. Abbot had placed her precious eggs where she did. If her son was to be left to find his own food, she would at least see to it that he was put where he would not starve for lack of the proper kind.

凯弟太忙了,忙着吃饭,忙着搭建自己的小木屋,根本顾不上关心别的事情,所以,和当初父母离开他的时候一样,孤身一人,没人烦他。想想这么小的一个小家伙出生还不到一天,就开始自己找食,随时准备张嘴吃第一口饭,是不是有点儿不可思议?可是,他一从卵壳里探出头来,就爬到一边,吃起了美味爽口的冬青叶色拉。这色拉要比麦乳精好吃多了,可以肯定,这就是当初袋袋蛾夫人把她的宝贝卵产在冬青树上的初衷。如果儿子得自己找食吃，那么，她至少要保证选的地方，不会让宝贝儿因为缺少合适的食物而挨饿。

That was all very well so far as his meals went; but where was the carpenter to be found who would make him a snug cabin? For, after he crept out of the egg-shell, he had no covering for his tender little body, and no son of the House of Abbot would be caught without just the right sort of shelter over his head. No, indeed. His family for many, many years back had all had tidy places of their own, and it was not to be thought for a moment that he would get along without one as good as any of them had had. He might be fatherless and motherless, and with no nurse to prepare his dinner for him; but homeless he would not be. Never! And he didn't sit round waiting for that carpenter, either. He went to work as busy as a little bee and made his own log cabin. That is what Keti did. Think of it, —with no one to show him how, —and he a baby just new in the world! He started it with the tiniest pieces, for he was wee himself and did not need a big dwelling; and if you had seen him turning heels over head with it in the beginning, you might

have thought that he was only playing a game with a cunning little collar. But he kept at it until it covered him all up, and as fast as he grew, he kept chopping more tiny holly logs and making it bigger, to fit him.

到目前为止,吃的是不用愁了。可是,到哪里去找木匠,为他搭建舒适温暖的小木屋呢?刚从卵壳里爬出来时,他弱不禁风,而且没有遮身之所。再说了,袋袋蛾家的儿子们可不能没有像样的遮身之所,绝对不能!许多许多年之前,他的祖先们都有属于自己的住所,而且都是干干净净、整整齐齐的。他的生活中会没有祖先那样像样的住所?这个念头连想都不要想。因为,他可以没有爸爸、没有妈妈,也可以没有保姆为他做饭,但是,他不能没有家。绝对不能!他也不能坐等木匠上门。于是,他像蜜蜂一样忙忙碌碌地工作,搭建起自己的小木屋。这就是凯弟干的事情。想想看,不但没人教他,而且他本身还只是一个刚出生的小不点儿!他先从小处着手,因为,他就不丁点儿大,不需要住多大面积。刚开始,看到他用脚后跟把正在搭建的小屋举过头顶时,你肯定会以为他是在玩一个精巧的小项圈。可是,他不停地玩着,最后小项圈竟然把他整个儿都遮住了。他发育的速度很快,可是,他以同样快的速度砍下更多的冬青树碎片,不断扩大他的“项圈”,总能把自己遮住。

He cut the wee twig-logs as neatly as a man could cut huge ones with an axe, although he had no choppers to use but his own strong jaws. He placed them criss-cross at the ends, and fastened them together firmly, not with wooden pegs or nails, but with silk.

他砍小树枝时,跟人类用斧子砍大树干一样麻利。不过,除了发达的上下颌之外,他可没有斧子可用。他把砍下来的小木条首尾交叉,牢牢地固定在一起,不过不是用木栓或木钉,而是用丝。

Now, no man who is not very rich has his walls covered with silk, for it costs a great deal to buy it. But it was none too good for Keti Abbot, who would have as fine a cabin as his father and his grandfather and his great-grandfather had had, even if he did have to make every bit himself. You will wonder where he got his silk. It grew in a sort of pocket in his body back of his head, and all he had to do when he was ready to spin was to pull it out from an opening in his mouth and swing it back and forth until he wove a silken lining for his little log cabin of a nest. And no one taught him how to spin or how to weave. He just did it all right the first time he tried.

要知道,只有富人才会在墙上贴丝,因为买丝得花很多钱。可是,对袋袋蛾凯弟来说,在墙上贴上丝算不了什么。袋袋蛾凯弟一心要把小屋搭得像他的爸爸、他的爷爷和他的祖爷爷的小屋那么精美。所以,就算是所有的活真的都得由他自己一点一滴地完成,他也在所不辞。那么,他的丝都是从哪里弄来的呢?是从他脑袋后面一个口袋模样的地方长出来的。准备吐丝的时候,他只需把丝从嘴巴里的一个小洞里拽出来,来回摆动,织成一片柔软光洁的衬里,衬在他的木屋小巢里面。没人教他怎么吐丝、怎么编织,他完全是一试就会。

After he had taken all that pains with his home, you will not be surprised to learn that he liked it so well that he stayed in it night and day, never leaving it for a minute all the fall. He no more left that little log case of his than a snail would leave her shell, and in some ways he got along in much the same way that a snail does. When he went walking about on the highroads of the holly branches, he stuck his head and creeping feet out of the open doorway and stepped off—cabin and all. While he was tiny, he walked on top

He travelled on the under side of the branch and let his home hang down like a bag.
他就在树枝下侧到处转悠,让他的房子像一个袋子那样倒挂起来。

of the branch, with his house straight up in the air; but after he grew and his house was too heavy to hold that way, he traveled on the under-side of the branch and let his home hang down like a bag.

修建小屋让他历尽了千辛万苦,知道这些,你就不会惊奇他居然如此喜欢他的小屋,竟然白天晚上都待在里面,整个秋天连一分钟都没离开过。正如蜗牛寸步不离自己的壳一样,袋袋蛾凯弟也寸步不离自己的小木屋。在有些方面,他和蜗牛的生活方式几乎一样。在冬青树的树枝上面比较平坦的地方四处转悠的时候,他会把脑袋和爬脚伸到门外的开阔处,结果一脚踩空——连屋带他整个掉了下去。小时候,他在树枝上面走动,整个房子都直竖在空中。可是,等他长大了,房子就太重了,没办法直竖起来,于是,他就在树枝下侧到处转悠,让他的房子像一个袋子那样倒挂起来。

In this way he would hunt here and there about the tree for the best holly salads; and having his dwelling handy by, he could camp out wherever he happened to be.

他会以这样的方式在树上四处寻找最好吃的冬青叶色拉。由于他总是随身携带自己的小屋,所以,他走到哪里就能在哪里安营扎寨。

Now all these things that Keti did without any practice—like catching his own salad at just the right stage, and cutting holly-twigs just the right length, and fitting them together at the corners in just the right way, and binding them into a snug little home with a silken lining which he spun and wove just right—are very wonderful things indeed. And perhaps it is not quite true to say that he was not taught, because, after all, he had a better teacher than can be found in all the colleges of the world, or even in all the kindergartens. Of course that famous teacher is Dame Nature herself; and the little chap got along so well with all the hard lessons of his life by simply obeying his instincts.

It isn't much use to pucker up your brow and try to understand how Dame Nature led Keti from task to task, and how he could do all these things perfectly the first time trying. You might study about it for more than one hundred years, and even then not understand very well. It is enough for us to know that Keti ate and cut and builded and spun and wove; so much any of us can find out for ourselves by some day watching one of his kind; for Keti grew up in time, and had children and

grandchildren and great-grandchildren; and although he never saw any of them himself, that is no reason why we cannot see them.

要知道,凯弟做的所有这些事情都是没经过任何训练的。比如，不早不晚刚好赶上吃自己的美味色拉;比如,砍下的冬青树枝不长不短正合适;比如,把这些树枝不偏不倚、正正好好地固定在屋角;再比如,把这些树枝捆扎成一个舒舒服服、暖暖和和的小屋,还在小屋的内墙贴上他自己吐丝织成的柔软光洁的衬里,不大不小正合适。所有这些事情真是太神奇了!也许,说他是无师自通不怎么准确。毕竟,他有一位老师,世上所有大学、甚至所有幼儿园里,都找不到比这位老师更好的了。当然了,这位杰出的老师就是大自然夫人本人。面对重重困难,这个小家伙能如此应付自如,凭的就是自己的本能。想知道大自然夫人到底是怎样引导凯弟完成了一个又一个任务的吗?想知道为什么所有这些事情凯弟都一试就会吗?告诉你没用。就算你紧锁眉头、冥思苦想也没用。你可以用一百多年的时间去寻根究底,但是,即使到了那个时候,你也还会百思不得其解。我们只要知道凯弟吃树叶、砍树枝、自己建房、吐丝贴墙就行了。某一天，无论我们谁观察到一只他的同类，也只能发现这么多了,因为时间一到,凯弟自然就会长大，然后，生儿育女，子子孙孙不断繁衍下去。虽然他自己从来也没见过他的子孙后代，但是，我们不可能见不到他们。

It was during the fall, when Keti was only a few weeks old, that the cardinal bird sat in the holly tree now and then, adding his bright feathers to the bright berries on the twig and making altogether a lovely picture, though Keti never noticed the gay visitor when he poked his head out of his cabin door.

红衣主教鸟时不时地来到冬青树上的时候，正值秋天。那个时候凯弟出生才几个星期。冬青枝头那一颗颗红果鲜艳欲滴，配上鸟儿那身亮丽的羽毛，简直就是一幅美丽的图画！不过，凯弟从屋门探出小脑袋的时候，从来也没留意过这位衣着艳丽的客人。

Neither did he see those little cousins of his on the trees near by, who were making other kinds of cases to live in. They were not such pretty ones as Keti's, and the sticks went up and down on the outside instead of crosswise, and perhaps they looked more like bags than log cabins. Still they did as well for shelter, and were as

Keti's cousins
凯弟的表亲们

carefully lined with silk. One was as large as Keti's home, but the others were much smaller. Each made his house in his own way, and never looked to see what his cousin was doing.

他也没有留意过旁边树上的那些同辈兄弟姐妹们。他们也在建房子,可是那些房子都不一样。他们的房子不如凯弟的好看,房子外面不是纵横交错的枝条,而是直上直下的。也许,他们的房子看起来不像是小木屋,倒更像是一个个小袋子。不过,这些袋子照样可以居住,房子里面也衬着一层密密麻麻的丝。其中有一个房子和凯弟的房子一样大,但是,其他的房子就小多了。他们搭建房子的时候都各行其是、随心所欲,从来也不看周围的同辈兄弟姐妹们是怎么建房子的。

So Keti worked at building and spinning day by day; and he walked here and there along the holly highroad, clinging to the under side of the branch, whenever he felt hungry enough to hunt up a good fresh dinner.

所以,凯弟日复一日地忙着建房子、吐丝;什么时候饿了,想吃一顿丰盛的新鲜饭菜了,他就会紧紧地抓住树枝下侧,沿着冬青枝高速公路到处转悠。

Well, after a while the weather grew colder—as cold, indeed, as it ever gets in the land where Keti lived; for winter came on and the little fellow, still less than four months old, was growing drowsy. So he fastened his cabin to a twig with silk, and swung there like a bird in a nest while he took rather a long nap. He was safe as a baby oriole in his swinging house, which rocked in the wind like the cradle old Mother Goose once sang about.

哎呀，没过多久，天冷了——的确很冷，当初凯弟待在地上的时候，也这么冷过。已经进入冬季，这个还不到四个月大的小家伙有点昏昏欲睡了。于是，他吐丝把自己的小木屋牢牢地系在了一根小树枝上，开始睡觉，这一觉可真不短。这期间，他就像巢中的鸟儿那样悬在半空中。他的房子像鹅妈妈歌中曾经唱过的那个摇篮一样在风中荡来荡去，可是，他很安全，安全得就像歌中摇篮里的黄鹂鸟宝宝一样。

And after he had slept there a number of weeks, three children ran along one day, looking for holly branches to make their house bright at Christmastime. Little Eleanor spied Keti's cabin swinging like a cradle and said, " Oh my! oh my! oh me! oh my! Here's

a tiny wee bag of sticks! I want it for Dolly Jane's very own little Christmas-tree. And there is a holly-berry stuck right in the side of it!"

有一天,三个孩子跑过来了。那时,他已经在这儿睡了好多个星期了。三个孩子是来找冬青树枝的,他们要在圣诞节把房子装点得漂漂亮亮、亮亮堂堂的。小埃兰娜发现了凯弟的小屋，它正像摇篮那样轻轻地摆动着。"噢,天哪！天哪！天哪！我的天哪！瞧,这一袋小树枝！太可爱了！我要把它拿回家，送给我的洋娃娃简儿,让她也有自己的小圣诞树。瞧，边上还挂着一颗冬青果呢！"

So David took out his knife and cut off Keti's cabin, twig and all; and Eleanor tucked it among the branches of a small evergreen tree which she was making ready for Dolly Jane's Christmas.

于是,大卫掏出他的小刀,连树枝带小屋都割了下来。埃兰娜把它塞进了一棵小常青树的树枝中间。这棵常青树就是她的洋娃娃简的圣诞树,她正忙着装点这棵圣诞树呢。

There Keti stayed while Eleanor and David and Phoebe helped make laurel wreaths, and while Father put up the big red balls, and while Mother hurried about making plum pudding and wonderful cakes and candies.

凯弟在那儿待着的时候，埃兰娜、大卫和菲比正在帮着做圣诞花环，爸爸在挂大红球，妈妈也在匆匆忙忙地准备葡萄干布丁、好吃的蛋糕和糖果。

On December twenty-fourth the three children crept away to bed very early, so as to be sure not to be too sleepy to waken if they should hear anything that sounded like reindeer stamping on the roof or any jingle that might be sleigh-bells; for Mother always read to them, —

'Twas the night before Christmas,
When all through the house
Not a creature was stirring, —
Not even a mouse, —

and the rest of those jolly verses, every Christmas Eve; and every time they could not help thinking how amusing it would be to see for themselves what really happened in their own home on that night so full of secrets.

12月24日晚上，三个孩子早早就爬到床上睡觉了，他们得保证听到驯鹿脚踩屋顶的声音或者雪橇叮叮当当的铃声的时候，不能困得醒不过来。每个平安夜，妈妈都要给他们读这个歌谣——

那是圣诞前夜

整个屋子里一片寂静

没有任何声音，

就连老鼠都静悄悄的——

还有其他一样好听的歌谣。每次听这些歌谣的时候，他们都禁不住要想：如果能亲眼看到那个神秘莫测的夜晚自己家里发生的一切，那该多有趣呀！

But it was not Eleanor or David or Phoebe who wakened that Christmas Eve. It was little Keti, who crept off Dolly Jane's tree right in the midst of all the fun. The house was so warm that he got over being drowsy, and suddenly the electric lights were turned on, which made the room bright as day.

可是，那个平安夜，醒来的不是埃兰娜，不是大卫，也不是菲比，而是小凯弟。大家正热闹的时候，他从洋娃娃简的圣诞树里爬了出来。他的小屋很暖和，所以，他醒了。突然，不知谁打开了电灯，把房间照得如同白昼一般。

And there was Somebody putting glitter all over the big Christmas tree until it sparkled like stars and icicles and shiny snow. And there were bright toy birds perched on the twigs, and ropes of gayest red and pretty gold and silver hanging all about the spreading green branches. And all around the base of the tree was a long line of toy animals like every kind you can see at the circus. A little reindeer led them, and after him came all the others, like a grand parade.

有个人正忙着往大圣诞树上装点光亮，最后，圣诞树开始闪闪发光，像闪闪烁烁的星星，像晶莹剔透的冰柱，也像光洁如玉的雪花。圣诞树的枝头挂着明晃晃的玩具鸟。舒展的绿色枝条上系满了鲜艳的红绳子和好看的金绳、银绳。树的底座四周是一长溜动物玩具，你在马戏表演里看到的那些动物应有尽有。领头的是一个小驯鹿，其他动物跟在他身后。看上去简直就像一支壮观的游行队伍。

And Somebody was cramming the stockings by

the fireplace so full of little packages that they were all bunched out. And there were toys and books and bundles tied with red ribbon in piles before the grate. And Somebody kept laughing softly and whispering, " Eleanor will love to swing Dolly Jane in this little hammock"; and " David will make wonderful things with this set of carpenter tools"; and "Phoebe will paint pictures in this book by the hour."

有个人正在往壁炉旁边的那些长筒袜里塞一小包一小包的礼物，塞得太满了，结果所有的长筒袜都鼓鼓囊囊的。炉栅前面是一堆一堆的玩具、书和系着红丝带的小礼物。有个人不停地轻声笑着，还不停地嘀咕着，“埃兰娜肯定喜欢这个小吊床，她会迫不及待地把她的洋娃娃简儿放在里面摇来摇去”；“这套木工工具会让大卫做出各种奇妙的东西来的”；“菲比会在这个本子上画画，一连画几个小时。”

Think of it! All those things happening while the three children slept soundly, and little Keti Abbot awake all that delightful Christmas Eve! He even felt so lively that he crept out to the nearest holly-branch and nibbled a lunch at midnight.

想想看，发生这一切的时候，三个孩子正呼呼大睡；

And fly forth to seek out a wingless mate.
他向前飞去，去寻找一个没有翅膀的伴侣。

那个愉快的平安夜，只有小凯弟彻夜未眠！他甚至感到浑身充满了活力，所以，大半夜的时候，他竟然爬到最近的冬青树枝上细嚼慢咽地享用起了夜宵。

While he was doing this, Somebody saw him creeping about and chuckled, and then shook a finger at him and called him a long name that sounded like talking in Latin. ' Oiketicus abbotii," Somebody whispered, "don't tell what you have seen tonight. It's a secret, you know."

他吃得正香呢，有个人发现了他。看到他爬来爬去的样子，他扑哧一声笑了，指着凯弟，说了一长串名字，听起来好像是拉丁语："噢开帝俄斯·艾伯特"。那个人小声对凯弟说，"不要把你今晚看到的事情说出去。这是秘密，明白吗？"

And sure enough, Keti, the little Christmas guest of Dolly Jane, never told a soul about what went on that gay night, although he crept from place to place and feasted during the twelve days before the holly, with him on it, was taken out-doors again. Then he went to sleep until warm weather came, and wakened

him to eat and grow and change into a little winged creature like his father, and fly forth to seek out a wingless mate in a log cabin like the one his mother had once lived in.

当然了,尽管凯弟在那里待了十二天,这期间他还爬来爬去、大吃大喝,但是,这位洋娃娃简儿的圣诞小客人,从来也没有把那个喜气洋洋的夜晚发生的事情告诉过任何一个人。十二天之后,那棵冬青树连同凯弟又被拿到了室外。于是,他又开始睡觉,一直睡到天气变暖才醒过来。他又开始吃呀、长呀,最后变成了一个像他爸爸那样带翅膀的小家伙。他向前飞去,飞到小木屋去寻找他的伴侣。她没有翅膀,她的小木屋和他妈妈当年住过的小木屋一样。